Tough Choices

TOUGH CHOICES

---

*Facing the Challenge
of Food Scarcity*

Lester R. Brown

*The Worldwatch Environmental Alert Series
Linda Starke, Series Editor*

W • W • NORTON & COMPANY
NEW YORK   LONDON

First Edition

The WORLDWATCH INSTITUTE trademark is registered in the U.S. Patent and Trademark Office.

The views expressed are those of the author and do not necessarily represent those of the Worldwatch Institute; of its directors, officers, or staff; or of its funders.

The text of this book is composed in Plantin
with display set in Zapf Book Medium.
Composition by Worldwatch Institute; manufacturing by the Haddon Craftsmen, Inc.

ISBN 0-393-04048-8
ISBN 0-393-31573-8 (pbk)

W.W. Norton & Company, Inc., 500 Fifth Avenue, New York, N.Y. 10110
http://web.wwnorton.com
W.W. Norton & Company Ltd., 10 Coptic Street, London WC1A 1PU

1 2 3 4 5 6 7 8 9 0

 This book is printed on recycled paper

Contents

Acknowledgments

Producing a book in a matter of months and in the midst of a heavy travel schedule is not easy. Without the dedicated support of my executive assistant, Reah Janise Kauffman, including many weekends of work, this book could not have been completed in time. Indeed, without her, I would not have attempted it.

Linda Starke, who is the series editor for these Environmental Alert books as well as *State of the World* and *Vital Signs,* was also an enormous help. At W.W. Norton & Company, Iva Ashner did everything possible to facilitate the book's production.

Toni Nelson helped with the research for the book, compiling much of the data used in the analysis and offering suggestions along the way. Colleagues Christopher Flavin, William Mansfield, and Sandra

Postel read the manuscript in various stages and made helpful comments. Outside the Institute, Dana Dalrymple, with the U.S. Agency for International Development, helped maintain the proper perspective. Carl Haub, senior demographer at the Population Reference Bureau, was kind enough to review all the population-related material in the book.

And we are indebted to Robert Wallace, who has a long-standing interest in the world food situation, for a personal contribution to support the book. The Curtis and Edith Munson Foundation also provided specific funding for the book. We value the enthusiastic support of both Angel Cunningham and C. Wolcott Henry at Munson. Doing the book drew on a broad base of information gathering and earlier research that was funded by the Carolyn, Nathan Cummings, Geraldine R. Dodge, Ford, George Gund, William and Flora Hewlett, W. Alton Jones, John D. and Catherine T. MacArthur, Andrew W. Mellon, Rasmussen, Surdna, Turner, Wallace Genetic, Weeden, and Winslow foundations; The Pew Memorial Trust; the Rockefeller Financial Services, and the Prickett, Rockefeller Brothers, United Nations Population, and the Wallace Global funds.

In addition to W. W. Norton & Company, we are grateful to the publishers who bring our books out in some 30 other languages. This book is already scheduled to appear in many of them.

Lester R. Brown

Worldwatch Institute
1776 Massachusetts Ave., N.W.
Washington, D.C. 20036

July 1996

Editor's Note

Something important is happening to the world food supply. A combination of water scarcity, land hunger, severe heat and drought, and record growth in demand have conspired to pull the world's carryover grain stocks down to their lowest level ever and to raise wheat and corn prices to unimaginable highs. Something has got to give; some adjustments have to made. But at what price? And who will bear the brunt of the cost?

Tough Choices takes a sobering look at this rapidly changing world food situation. It is being published at a time when the international community and political leaders everywhere are taking a fresh look at the food prospect. More than 20 years ago, an earlier generation of people concerned about feeding the world gathered in Rome for the first World Food Conference, at which

Henry Kissinger was widely quoted as declaring that within a decade, no man, woman, or child would go to bed hungry. That goal was never achieved, and the chances of meeting it in the foreseeable future have worsened considerably since 1974.

As world leaders and the heads of U.N. agencies gather again in Rome for a conference on food prospects, they need to take a more realistic look at the situation. We hope that *Tough Choices* contributes to that process.

As Lester points out in his Foreword, this is the third book in the Environmental Alert Series to deal with food supplies and the pressures placed on them by growing populations and increasing affluence, which gives people the means to buy more food and more grain-intensive products. Other books in the series have considered water scarcity, the consumer society, and the coming energy revolution, among other topics. (See list of titles on page 2.) Next is *Fighting for Survival: Environmental Decline, Social Conflict, and the New Age of Insecurity*, by Michael Renner, also being published this fall. We look forward to hearing from you with any comments about these books or the Institute's two annual publications, *State of the World* and *Vital Signs*.

Linda Starke, Series Editor

Foreword

This book is one of our contributions to the World Food Summit being held in November 1996 in Rome. Other contributions include Worldwatch Papers by Gary Gardner, on the worldwide scarcity of cropland, and Sandra Postel, on water use in agriculture.

Tough Choices is also part of an ongoing debate between Worldwatch Institute and both the U.N. Food and Agriculture Organization (FAO) and the World Bank about the global food prospect over the next 15 years. Both FAO, which does the official global supply and demand projections, and the World Bank, the principal international development agency, are projecting a future of surpluses and declining grain prices.

Worldwatch Institute, in contrast, has been arguing that the issue is not going to be food surpluses but scarcity—not declining grain prices, but rising prices. If the world is indeed heading toward a future of scarcity, then projections indicating surpluses and declining grain prices are contributing to potentially serious underinvestment in agriculture and family planning.

Recent research at the Institute on this issue has been concentrated in three books. The first, *Full House: Reassessing the Earth's Population Carrying Capacity*, was published in August 1994. The second, *Who Will Feed China? Wake-Up Call for a Small Planet*, came out in October 1995. The third is this book, *Tough Choices: Facing the Challenge of Food Scarcity*, released in September 1996. All three argue that understanding the food prospect requires taking into account not only economic trends but also a broad range of environmental issues, including the sustainable yield of oceanic fisheries and of aquifers, the physiological capacity of crop varieties to use fertilizer, the cumulative effect of soil erosion on land productivity, and rising global temperatures.

Full House, written with my colleague Hal Kane, was our contribution to the September 1994 U.N. Conference on Population and Development held in Cairo. It analyzed the world food prospect, focusing on the relationship between projected growth in demand in key countries and their production potential. It was an overview of the major trends that are shaping the world food prospect. *Full House* focused on the projected imbalance between the growing demand for grain fueled by both population growth and rising affluence and the future growth in grain production as various constraints, most importantly water scarcity and a diminishing response of grain yields to fertilizer, begin to take effect.

The two years since *Full House* was written have confirmed the loss of momentum in efforts to expand the world food supply. There has been no growth in the fish catch, suggesting that the marine biologists are right—that the world is indeed pressing against the sustainable yield of oceanic fisheries. With grain production, the last two years have provided additional evidence that the growth in the world grain harvest is slowing.

In 1994, we projected that the food supply would tighten rather dramatically as we moved into the early twenty-first century. We did not at the time anticipate that grain prices would double between spring 1994, when *Full House* was being written and spring 1996. Under no circumstances did we expect to see wheat trading for $7 a bushel and corn for $5 a bushel during this century.

Full House argued that there was a need to reorder priorities. Most important was the need for a greater effort to slow population growth. The book strongly supported the goals of the Cairo conference to lower fertility by filling the family planning gap and investing heavily in the education of girls and young women.

Working on *Full House* reminded me of the potential enormity of the grain deficit looming in China. The combination of a population growing by 13 million per year and incomes rising at double-digit rates was driving up the demand for grain at an unprecedented rate. Meanwhile, on the supply side, the extraordinary rate of industrialization that made this record economic growth possible was leading to the wholesale conversion of cropland to nonfarm uses and to the diversion of irrigation water to the cities. The combination of these trends could lead to a potential grain deficit so massive that, on top of rising demand for imports in other developing countries, it could eventually overwhelm the export capacity of the United States and other exporting countries.

My concerns were first expressed in an article in the September/October 1994 issue of *World Watch* magazine. Later, they were developed in *Who Will Feed China? Wake-up Call for a Small Planet.* Within China, where officials had been impressed with the extraordinary grain output gains during the first several years following the economic reforms of 1978, there was a feeling that these trends would continue indefinitely. China's leaders failed to realize that the growth in the grain harvest by roughly half during the first six years after the reforms was largely a one-time gain.

Initially, officials in Beijing were strongly critical of my analyses, and they denied that China would ever import grain other than for seed. But as grain prices climbed nearly 60 percent there in 1994, forcing Beijing to import massive quantities in an attempt to stabilize prices and curb urban unrest, government attitudes changed. One manifestation of this change was an invitation I received in the spring of 1995 to visit Beijing. The resulting indepth discussions with officials from both the Ministry of Agriculture and the Chinese Academy of Sciences helped to define more sharply the issues that China is facing on the food front.

Overnight, China had become the world's second largest grain importer, trailing only Japan. Outside China, it gradually dawned on people that 1.2 billion Chinese moving up the food chain—consuming more pork, poultry, eggs, beef, beer, and other grain-intensive products—would alter the world's food balance, raising food prices everywhere. The realization that China's changing food situation would affect the entire world led to literally hundreds of conferences, seminars, studies, and new agricultural projections. Invitations for me to meet with senior management teams of agribusiness firms, farm leaders, commodity traders, investors, and others far exceeded my capacity to respond. Feeding China is now recognized, both within and outside China, as a major challenge for the entire world.

This book focuses on the policy response to scarcity— the kinds of choices that governments will have to make as they try to ensure the security of food supplies in the years ahead. The debate continues between FAO and the World Bank on the one hand and Worldwatch Institute on the other. But as of June 1996, there are some signs of change. After several months of meetings and dialogue with World Bank officials, including President James Wolfensohn, a letter from him to the Institute in early June indicated that the Bank would be releasing a new assessment of long-term food security prospects by September 15. Wolfensohn also indicated that the Bank would take into account the wider range of factors we have urged are needed for assessing the food prospect.

As yet there is no word from FAO, but at some point the organization will have to deal with the growing disparity between its projections since 1990 and actual trends. (See Chapter 2.)

On the positive side, there is a growing concern about food prospects among political leaders and ministers of both environment and agriculture. While in Europe in early 1996 to launch various language editions of *State of the World* in eight countries, I was invited to meet with three heads of state. All of them wanted to discuss the food prospect. Four things seem to be feeding their concerns. One is the sense that we may have hit the wall with oceanic fisheries even as world population continues to grow by nearly 90 million a year. Two, water scarcity is beginning to command the attention of political leaders everywhere, even in Europe, a relatively well watered region. Three, there is growing concern about the effect of global warming, and particularly of increasingly intense heat waves, on the food prospect. And four, the potential effect on world food prices of 1.2 billion Chinese consumers moving up the food chain and importing massive amounts of grain is beginning to sink in.

Although the growing imbalance between food and people is still far from getting the attention it deserves, the level of concern is rising. And, more important, we are beginning to realize that the steps needed to secure the food supply for the next generation are the same as those needed to build an environmentally sustainable economy.

Lester R. Brown

Tough Choices

1

The Challenge

The thesis of this book is that food scarcity will be the defining issue of the new era now unfolding, much as ideological conflict was the defining issue of the historical era that recently ended. Even more fundamentally, *Tough Choices* argues that rising food prices will be the first major economic indicator to show that the world economy is on an environmentally unsustainable path.

An early hint of the shift to an economy of scarcity came in late April 1996, when wheat prices on the Chicago Board of Trade soared above $7 a bushel, the highest level in history and more than double the price a year earlier. Corn was selling for more than $5 a bushel, also a record. And the price of rice, the third of the three grains that provide most of humanity's food and feed, was on the way up.[1]

Prices were climbing because world carryover stocks of grain had fallen to 48 days of consumption, the lowest level on record. As grain futures prices surged in the late spring, news coverage moved from the financial pages of newspapers to the front pages, reflecting an urbanized society's deepening concern about the security of its food supply.[2]

Even though we live in an age of high technology—space exploration, the World Wide Web, and genetic engineering—humanity was suddenly faced in 1996 with one of the most ancient of challenges: how to make it to the next harvest. In the popular media, this worrying turn of events was explained largely by drought in the winter wheat region of the Great Plains. But was weather the cause of the runaway price rises? Or was it the precipitating event, bringing a deteriorating situation into focus? Was the dramatic climb in prices in late 1995 and early 1996 a temporary matter, a blip on the price charts? Or did it signal a fundamental shift from a world food economy dominated by surpluses to one that would be dominated by scarcity?

The World Bank and the U.N. Food and Agriculture Organization (FAO), which does the projections of world grain supply and demand that governments rely on, both argued that this was indeed a temporary situation, and that the issue in the future would be surplus capacity and declining grain prices. This book argues the opposite—that the issue will be scarcity and rising grain prices, that the future of the world food economy will not be a simple extrapolation of the past.[3]

Although grain prices will fluctuate in the years ahead, and will decline in the short run from the record highs reached in the spring of 1996, the thesis of this book is that the scarcity of early 1996 was not a fleeting phenomenon but the result of the collision between continually expanding human demand for food and some of the earth's natural limits, including the sustainable yield of oceanic fisheries and of the aquifers that supply irrigation water, and the physiological limits of crop varieties to use fertilizer. *Tough Choices* also warns that there is no identifiable technology waiting in the wings that will lead to a quantum jump in food production comparable to those that came from earlier technological gains, such as the discovery of fertilizer or the hybridization of corn.

It has been clear to many analysts for some time that the trends of environmental destruction of the last few decades could not persist without eventually undermining the economy. Continuing deforestation will lead to increased runoff, more destructive flooding, and greater soil erosion. Heavy losses of topsoil from erosion will lower land productivity, threatening the food security of the next generation. Similarly, satisfying the growing demand for water by overpumping aquifers will one day lead to sharp reductions in water supplies when aquifers are depleted. If we keep adding excessive amounts of carbon dioxide to the atmosphere, we face the prospect of higher temperatures and more extreme weather, including more destructive storms and floods, more severe droughts, and more intense heat waves.

If we continue to overfish, still more fisheries will collapse, leading to rises in seafood prices that will make those of recent years seem modest by comparison. A continuing release of chlorofluorocarbons (CFCs) into the atmosphere will deplete the stratospheric ozone layer that protects not only us but all life on earth from harmful ultraviolet radiation. And if plant and animal species are extinguished at the rate of the last few decades, we face a massive biological impoverishment, one that will rival that associated with the cataclysmic event that eradicated dinosaurs and much of life some 65 million years ago.

If we keep adding nearly 90 million people to our numbers each year, all of whom want to become affluent, our life-support systems will be overwhelmed. The persistent release of pollutants into the air and water will not only lead to human health problems, it will damage crops and disrupt the functioning of ecosystems. All these environmentally destructive trends adversely affect the food prospect.[4]

Although there is public recognition of the need to reverse these trends, only one has been reversed to date. The one success came with CFCs after scientists discovered the hole in the ozone layer over Antarctica, a finding that scared politicians into action. This led to a highly successful international effort that reduced the manufacture of CFCs by 77 percent between 1988, the historical high, and 1995.[5]

Thus far, the political process has failed to respond to the urging of scientists and environmentalists to reverse the rest of the obviously threatening trends. Many people in turn have despaired of creating a sustainable global economy, concluding that it would take some dramatic event, perhaps a human-induced catastrophe of some sort, to mobilize societies. Some thought the wake-up call would come from an epidemic level of environmentally induced illnesses, including soaring rates of cancer. Others expected the loss of species, leading to ecosystem collapse, would be the turning point.

This book argues that it will be food scarcity that rouses us from our sleepwalk through history, convincing us to take the steps needed to create a sustainable balance between ourselves and our natural support systems.

Most environment ministers know the litany of environmental failures just described. They are all too familiar with the trends that need to be reversed. They agree on the need to shift the tax burden from personal and corporate incomes to environmentally destructive activities, such as carbon emissions or the generation of toxic wastes. They see the importance of taxing the use of virgin raw materials, in order to convert today's throwaway economy into a reuse-recycle one. And they know that the time has come to abandon the tax incentives that governments provide to couples for having children.

They also know this will not be easy. The road to an environmentally sustainable, food-secure future is paved with tough choices—choices that food scarcity will bring into focus. The food security of the next generation now depends on investing heavily in family planning, on educating girls in developing countries, and on creating equal opportunities for women. It means restructuring the energy economy, shifting from fossil fuels to renewable resources. It requires investing far more in energy efficiency. And it means pricing water at a level that recognizes its scarcity and raises the efficiency with which it is used.

Political leaders are now wrestling with food scarcity and struggling to grasp the dimensions of the new challenge posed by the recent reversals on the food front. What happened to the steady rise in grain production per person that the world had enjoyed for a generation? What happened to the trends that doubled seafood consumption per person between 1950 and 1990? What happened to the food-secure world that we were planning to leave our children?[6]

With the amount of grain in the bin at the start of the

1996 harvest at the lowest level on record, there is an urgent need to rebuild stocks, to provide at least a minimum of food security. But the addition of nearly 90 million people to the world each year and the rise in affluence in Asia make it more difficult to accomplish this. Early estimates released by the U.S. Department of Agriculture in June indicate that the 1996 harvest, assuming normal weather, will be barely enough to cover demand—allowing little to rebuild depleted stocks. This leaves the world in a high-risk situation.[7]

In 1988, when severe heat and drought reduced the harvest below consumption, the United States was able to satisfy its export commitments to more than 100 countries only by drawing down its stocks. Now, with almost no stocks to draw down, that would not be possible. If the United States were to have another poor harvest, like the one in 1988, it would lead to chaos in world grain markets by fall.[8]

By far the greatest threat to future food security is population growth. Those of us born before 1950 belong to the first generation ever to witness a doubling of world population. Stated otherwise, there has been more population growth since 1950 than during the preceding 4 million years—the period since our early ancestors first stood upright. Throughout most of our existence as a species, our numbers were measured in the thousands; today they are measured in the billions. The environmental effect of this enormous growth still eludes us.[9]

As the annual additions to world population have increased in recent decades, human demands have begun to overwhelm local life-support systems. Nowhere is this more evident than in oceanic fisheries. From 1950 to 1989, the oceanic fish catch climbed from

19 million to 89 million tons. This fourfold growth doubled the catch per person from 8 to 17 kilograms.[10]

Since 1989, however, there has been no growth in the oceanic fish catch, and the catch per person has fallen by 11 percent. FAO marine biologists report that all 15 oceanic fisheries are now being fished at or beyond capacity; 13 are in a state of decline. Our failure to stabilize world population size before reaching the limits of oceanic fisheries means that we now face a shrinkage in the seafood catch per person and continually rising prices as long as population growth continues. Nothing better illustrates the contrast between the old era, roughly the period from 1950 to 1990, and the new one than these figures on per capita seafood catch.[11]

Troubling new trends are emerging with land-based food production as well. After nearly tripling from 1950 to 1990, the world grain harvest has increased little since the bumper harvest of 1990. Early estimates of the 1996 harvest at 1.83 billion tons show a gain of less than 3 percent over the 1990 harvest of 1.78 billion tons. With each passing year the accumulating evidence suggests that it is becoming more difficult to sustain rapid growth in the grain harvest. (See Part II.)[12]

From the beginning of agriculture until mid-century, gains in production came primarily from expanding the area under cultivation. By 1950, the frontiers of agricultural settlement had largely disappeared and farmers quickly shifted to raising the productivity of the existing cropland, drawing on a backlog of unused technology that had accumulated over the preceding century. Now the world's farmers are encountering difficulties on that front. After more than doubling from 1950 to 1990, grain yield per hectare has increased much more slowly since 1990.[13]

In addition to facing a scarcity of productive new land to plow, farmers must cope with water scarcity as well. Aquifer depletion and the diversion of irrigation water to cities are diminishing the water available for irrigation in many countries. Water tables are falling in the key food-producing regions, including the southern Great Plains of the United States, several states in India, and much of northern China.[14]

Some countries will soon face abrupt reductions in irrigation water supplies as aquifers are depleted. At that point, the amount of water pumped is necessarily reduced to the rate of aquifer recharge. Irrigation cutbacks from overpumping renewable aquifers will have the greatest effect on food production in China and India, which rank first and third in world grain production (the United States is second) and which together contain two fifths of the world's irrigated land.[15]

Even more serious over the longer term is the decline in irrigation in regions dependent on what are essentially fossil aquifers, those that have little or no recharge. This includes the southern Great Plains of the United States, much of the wheat-growing region in Saudi Arabia, and parts of Libya and Tunisia.[16]

In the United States, depletion of the fossil aquifer under the southern Great Plains reduced the irrigated area in Texas, a leading U.S. agricultural state, by some 11 percent between 1982 and 1992. In Saudi Arabia, where exploitation of a deep fossil aquifer depended on a support price for wheat that was four times the world market level, the cutback has been more dramatic. As the program has become too costly for the fiscally stressed Saudi government to cover, irrigation cutbacks reduced the wheat harvest from 4.1 million tons in 1992 to an estimated 1.3 million tons in 1996.[17]

As countries begin to press against the limits of their water supplies, the competition between the countryside and cities intensifies. In this battle, cities almost always win. As irrigation water is diverted from agriculture, as in China, grain imports rise. Importing a ton of wheat is equivalent to importing a thousand tons of water. In effect, grain becomes the currency with which countries balance their water books.[18]

While many people are aware of pressures on the sustainable yield of fisheries and aquifers, few know about an even more troubling limit—the physiological capacity of existing crop varieties to use fertilizer. Between 1950 and 1989, farmers increased their use of fertilizer almost every year, boosting it from 14 million to 146 million tons. But as application rates climbed, diminishing returns eventually set in. The growth of fertilizer use has levelled off or even declined somewhat in North America, Europe, the former Soviet Union, and Japan. Fertilizer use in China may also be starting to level off.[19]

The engine that drove the near tripling of the world grain harvest from 1950 to 1990 is losing steam. The principal explanation for the lack of growth since 1990 has been the decline in world fertilizer use that began in that year. The old formula of combining more and more fertilizer with ever higher yielding varieties to expand the grain harvest is no longer working very well. Unless agricultural scientists can quickly find a new formula, the world is almost certain to face politically destabilizing food shortages in the not too distant future.[20]

Farmers have always had to deal with the vagaries of weather, but now they also must face the effects of climate change. As greenhouse gases build in the atmosphere, the earth's temperature rises. The 11 warmest years since recordkeeping began in 1866 have all occurred since 1979, with the three warmest years coming in the nineties.[21]

Crop-withering heat waves across the northern tier of industrial countries shrank the 1995 world grain harvest, making it the smallest since 1988. Even if all land set aside under commodity programs had been in production in 1995, carryover stocks of grain still would have declined. Within the United States, intense heat and drought have reduced markedly three of the last eight harvests. As noted earlier, the 1988 U.S. grain harvest dropped below consumption for the first time in history, something most analysts never expected to see.[22]

While the growth in world grain production is losing momentum, the growth in demand is on the upswing. Although some of this is due to nearly 90 million people added to world population each year, it is the rising affluence in Asia, enabling people to move up the food chain and to consume more livestock products, that is putting heavy new pressure on the world's grain supplies. Over the last four years, China has been the regional pacesetter. Its economy expanded 12 percent in 1992, 14 percent in 1993, 11 percent in 1994, and just over 10 percent in 1995—for a total of 57 percent in four years. The income of 1.2 billion Chinese has increased by more than half in four years. Much of this additional income was used to diversify diets by moving up the food chain, consuming more pork, poultry, eggs, beef, and other grain-intensive products.[23]

The record economic growth that began in Japan a generation ago and that spread to South Korea, Taiwan, Hong Kong, and Singapore has reached China, the world's most populous country, and is now spreading throughout Asia. The International Monetary Fund reports that the Asian economy (excluding Japan) expanded by over 8 percent in 1995, the third consecutive year of 8-percent growth for the region, and projects that it will grow by a similar rate in 1996.[24]

There is no precedent for the rise in affluence of so many people. When Western Europe embarked at midcentury on a period of rapid economic growth and the associated rise in consumption of livestock products, it contained 278 million people. The United States and Canada together at that time had 166 million people. As the economies in these two regions expanded rapidly, enabling people to move up the food chain, their claims on world grain supplies increased sharply. But the scale of this movement pales when compared with the 3.1 billion in Asia who are now setting off on a similar economic journey and at a much faster pace. Neither Western Europe nor North America ever came close to sustaining an economic growth rate of 8 percent.[25]

Rising affluence in land-scarce Asia is beginning to alter the pattern of world grain trade. China, for example, is emerging as a massive importer. In trade year 1994, China was a net exporter of 8 million tons, mostly corn. In 1995, it became a net importer of 16 million tons, mostly wheat. Overnight it emerged as the world's second largest importer, trailing only Japan. The effect on the world grain balance of this 24-million-ton shift in just one year in China's grain trade is comparable to what would happen if Canada, the world's number two exporter, suddenly disappeared from the face of the earth.[26]

Although world grain production did not expand from 1990 to 1995, the world added 450 million people during this five-year span, leading to the record drawdown in carryover stocks of grain mentioned earlier. With consumption exceeding production in four of the last five years, 1996 carryover stocks fell to the lowest level on record. This drawdown, leaving little more than pipeline supplies, is a one-time event. It cannot be repeated, which leaves rising prices as the only way to correct future demand/supply imbalances.[27]

The principal effect of rising food prices on the affluent will be indirect. In the United States, for example, the wheat in a $1 loaf of bread likely costs no more than 6¢. So even if the price of wheat doubled, that of the loaf of bread would only increase to $1.06. For livestock products, the effect would be somewhat greater because the cost of feed figures more prominently in the overall production costs. Nonetheless, since the U.S. farmer gets less than a quarter of the consumer's food dollar, the effect of rising commodity prices on the consumer is muted by the heavy food-processing industry expenditures on transportation, processing, and marketing. Although it might be inconvenient to spend a somewhat larger percentage of income on food, price rises are not life-threatening for the affluent.[28]

The same cannot be said, however, for poor consumers, who may already be spending 70 percent of their income on food. Holding their governments responsible for food shortages, they can be expected to take to the streets if they do not get help in the form of food aid or subsidized food. If the world's farmers cannot restore the rapid growth in the grain harvest that prevailed before 1990, urban food riots could become commonplace. Rising food prices may also drive unprecedented numbers of hungry people across national borders.[29]

But the shift in the world grain market is not simply about scarcity that will lead to hunger and nutritional deprivation among the poor, but more broadly about political stability. Since no economic indicator is more politically sensitive than food prices, scarcity raises the specter of political instability on a scale that could undermine economic progress. In an integrated world economy, rising food prices in one major country can quickly become rising food prices everywhere. When grain prices in China climbed nearly 60 percent in 1994, the government, fearing food riots in the cities, imported massive quantities of grain in an effort to check the price rises, thus triggering a rise in world grain prices in 1995 and 1996.[30]

Rising food prices in grain-importing countries, such as China, Egypt, and Mexico, could create potentially unmanageable problems. Political instability in any one of these countries would affect every facet of the global economy. In today's world, political instability anywhere can affect profit-and-loss statements of multinational corporations everywhere. The economic stability of individual corporations, the performance of stock markets, and the earnings of pension funds could all be affected by political instability in a key country.

There are few things that the government in Beijing fears more than the political instability that could accompany rising food prices. In Egypt, similar increases could destabilize the government, affecting political stability throughout the Middle East; in a country that imports more than one third of its grain, and whose population is expected to nearly double by 2030, rising grain prices are already exacerbating the balance-of-trade deficit. And in Mexico, which is importing growing quantities of wheat as water scarcity impedes its own agricultural expansion, the problems could be multidimensional. They could further weaken the value of the peso. They could also lead to record numbers of desperately hungry people crossing the border into the

United States in search of food.[31]

In extreme cases, food scarcity may lead to political instability and social disintegration. Ethnic conflicts could be exacerbated. Social disintegration of the sort afflicting Afghanistan, Somalia, Liberia, Rwanda, and Haiti may begin to spread to other countries.

The world is now moving into uncharted territory on the food front, facing problems that dwarf those of the past. Scores of governments in developing countries are faced with tough choices, including many they have never had to face before. Without a massive mobilization by governments to stabilize populations sooner rather than later, future economic and political stability cannot be assured.

Governments in developing countries, where the growth in population is already exceeding the food carrying capacity of local life-support systems, are faced with particularly tough choices in population policy. Those that have waited too long to slow their population growth may have to choose between the reproductive rights of the current generation and the survival rights of the next generation. Governments everywhere will have to make difficult decisions in land use, water use, climate policy, and the policies that deal with equity—how resources that are no longer expanding are distributed within populations that are still growing.

Farmers alone can no longer balance food supply and demand. Investing much more in agriculture and agricultural research is necessary, but it is not sufficient. Achieving an acceptable balance between food and people may now depend more on family planners than on fishers and farmers. Decisions made in the Ministries of Energy that affect climate stability may more directly affect the food security of the next generation than those made in the Ministries of Agriculture.

If the analysis in *Tough Choices* is at all close to the mark, food scarcity will provide the wake-up call the world has long needed. At some point, uncontrollable rises in food prices will provide clear evidence of excessive pressure on fisheries, aquifers, and croplands and of climate disruption. The years immediately ahead may provide the last chance to reverse the demographic and the environmental trends that are undermining the future of many countries before political instability and social disintegration make economic progress difficult if not impossible.

2

The Debate

At the beginning of this decade, there were two widely diverging views of the world food prospect. The World Bank and the U.N. Food and Agriculture Organization (FAO) argued that the problem would be surpluses and declining grain prices, based on their global grain supply and demand projections to 2010. Our own view was that growth in food production was slowing and that the problem would be scarcity and rising food prices.[1]

Economists at the Bank and FAO based their projection of surpluses on continuing advances in technology on a scale that would keep production running ahead of the growth in demand. Worldwatch—taking into account advances in technology, a shrinkage of productive new land to plow, the diminishing response to the use of additional fertilizer, the growing scarcity of fresh water, a heavy loss of cropland to nonfarm uses in the rapidly industrializing countries of Asia, the cumulative effects of soil erosion on land productivity, and the increasing frequency of crop-damaging heat waves associated with rising global temperature—argued that growth in output would slow, lagging behind the growth in demand. While Bank economists argue that extrapolation is the key to future trends, we do not believe that the future will be a simple extension of the past.

Now that we are in the sixth of the projected 20 years, we can begin to evaluate the trends. The bottom line is that the world has suffered a dramatic loss of momentum in the efforts to expand its food supply. The near tripling of the world grain harvest by the world's farmers between 1950 and 1990—from 631 million to 1,780 million tons—was a remarkable performance, one without historical precedent. But since the bumper crop of 1990, there has been little growth in the world grain harvest. (See Figure 2-1.) Early estimates of the 1996 harvest, assuming normal weather during the summer growing season, indicate a crop of 1.83 billion tons, less than 3 percent above the level in 1990.[2]

In addition to the broad-based trends just cited that are slowing growth in output, political and economic disruptions can also affect food production. Grain production has declined sharply in the former Soviet Union since 1990, dropping more than a third by 1995, largely as a result of various adjustments associated with economic reform. For example, fertilizer prices rose steeply when the subsidy that encouraged fertilizer use was discontinued, leading to a dramatic fall in usage. Large areas of marginal land that were cultivated prior to the economic reforms were rejected by the newly established market economy as unprofitable and were abandoned.

Million Tons

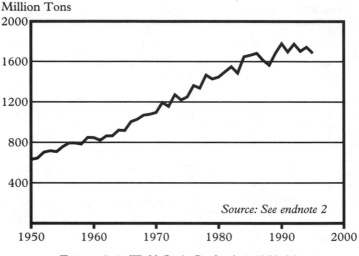

FIGURE 2–1. *World Grain Production, 1950–95*

In addition, the economic disruptions associated with the reforms and the weak prices for farm products resulting from the severely depressed state of the economy discouraged investment in agriculture and the use of inputs.[3]

One consequence of the slower growth in the world grain harvest in the late eighties and the limited growth since 1990 has been a precipitous decline in the grain harvested per person. After peaking at 346 kilograms in 1984, this number dropped to 295 kilograms per person in 1995, the lowest level since 1967. (See Figure 2-2.)[4]

Although there has been little growth in the world grain harvest since 1990, some 450 million people were added to world population between 1950 and 1995, as noted in Chapter 1. The continuing rise in the demand for grain was satisfied in large measure by drawing down grain stocks. Between 1991 and 1996, world carryover stocks of grain dropped from 339 million to 229 million

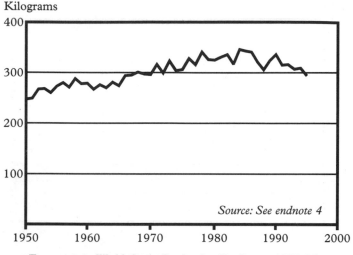

Kilograms

FIGURE 2-2. *World Grain Production Per Person, 1950–95*

tons. This reduction of 22 million tons per year dropped stocks to 48 days of world consumption, the lowest level in history. (See Figure 2-3.) Once carryover stocks drop below 60 days, prices become highly volatile. It is thus no surprise that world prices for wheat and corn doubled between early 1995 and mid-1996.[5]

Not only are stocks at the lowest level on record, but rebuilding them is going to be difficult. On the demand side, the world continues to add nearly 90 million people a year. And as noted in Chapter 1, a large share of humanity, most of it in Asia, is moving rapidly up the food chain, consuming more meat, milk, and eggs—all grain-intensive products.[6]

The second land-based food system, the world's rangelands, covers an area roughly double that of cropland. One tenth of the earth's land surface is suited to producing crops, but two tenths of it is used for grazing

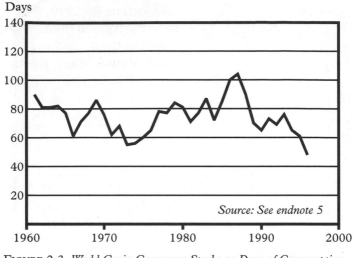

FIGURE 2-3. *World Grain Carryover Stocks as Days of Consumption, 1961–96*

cattle and sheep, producing most of the world's beef and mutton. The 3.2 billion cattle, sheep, and goats that supply the world's beef and mutton are ruminants, animals with complex digestive systems that enable them to convert grass or foliage from shrubs into meat and milk. In the pastoral economies of North Africa, East Africa, and South Africa, in the Middle East and Central Asia, and in large areas of the Indian subcontinent and China, range-based livestock provide not only milk and meat for hundreds of millions of people but a livelihood as well.[7]

Livestock grazing is the only practical means of producing food on this vast land area that is too dry, too steeply sloping, or too infertile to sustain crops. Unfortunately, human demands for livestock products in these regions are running up against the limits of rangelands. Overstocking with cattle, sheep, and goats is now the rule, not the exception. After expanding from 24 million tons in 1950 to 62 million tons in 1990, world production of beef and mutton has increased little. (See Figure 2-4.) Rangeland productivity cannot be enhanced by fertilization or other human interventions, given the arid nature of the resource base, but it can be reduced by overgrazing. Unfortunately, the carrying capacity of rangelands is deteriorating in much of the world.[8]

The third food system, oceanic fisheries, may also have reached its limit. From 1950 to 1990, the world fish catch increased from 19 million tons to more than 85 million tons, expanding more than fourfold. This spectacular growth—which doubled the world seafood catch per person, boosting it from 8 kilograms in 1950 to 16 kilograms in 1990—has apparently come to an end. (See Figure 2-5.) As noted in Chapter 1, FAO marine biologists report that all 15 oceanic fisheries are now being fished at or beyond capacity, and 13 are in a state

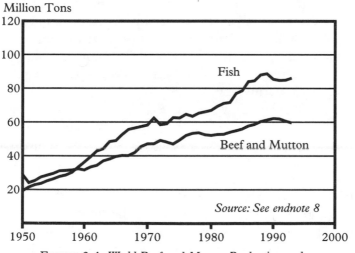

FIGURE 2-4. *World Beef and Mutton Production and Oceanic Fish Catch, 1950–93*

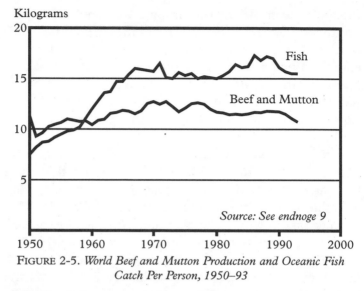

FIGURE 2-5. *World Beef and Mutton Production and Oceanic Fish Catch Per Person, 1950–93*

of decline.[9]

With oceanic fisheries, as with rangelands, human intervention cannot significantly boost productivity. But it can reduce it or, in some cases, even destroy the resource totally. The prospect of no growth in the production of beef or mutton or in the oceanic fish catch in a world that continues to add nearly 90 million people a year is a sobering one.[10]

Although the six years since 1990 do not make a new trend, they do raise concern about the ability of the world's farmers, ranchers, and fishers to expand the food supply rapidly in the decades ahead. With rangelands now being grazed at full capacity, augmenting the supply of beef and mutton depends on feedlots and requires more grain. Similarly, increasing the world fish supply now rests on fish farming, which in turn requires grain and soybean meal or other protein meal as feed. (In contrast to fish taken from the oceans, those produced in fish ponds or cages have to be fed.) Replacing the historical 2-million-ton annual growth in the seafood catch with either farm-raised fish or broilers would take a minimum of 4 million tons of additional grain a year, roughly the amount consumed in Belgium. The bottom line is that expanding supplies of animal protein—whether fish, beef, mutton, pork, poultry, or dairy—depends on increasing the supply of grain.[11]

This brings us back to the prospects for doing so. Official projections by the World Bank and FAO through 2010, which differ from each other only in minor details, indicate that we should be able to increase grain supplies rather dramatically. Using 1990 as a base, the World Bank assumes that the harvest will expand on average 26 million tons a year between 1990 and 2010, while FAO projects a slightly faster growth of nearly 28 million tons a year.[12]

Yet Bank projections have overestimated the grain harvest every year since 1990, with the discrepancy ranging from 56 million tons in 1992 to 225 million tons in 1995. FAO projections show an even greater gap. With each passing year, the difference between these two projections and reality grows wider. (See Figure 2-6.)[13]

Assuming that the lack of growth in the world grain harvest from 1990 to 1995 is a new trend would be as misguided as the Bank and FAO assuming that the trend from 1950 to 1990 will continue indefinitely. But the factors responsible for the lack of growth over the last five years need to be carefully examined to assess their likely effect on production over the longer term.

The World Bank is quite explicit about its reliance on extrapolation to determine future food production levels. In *The World Food Outlook*, the Bank study that contains its agricultural projections, the authors note that

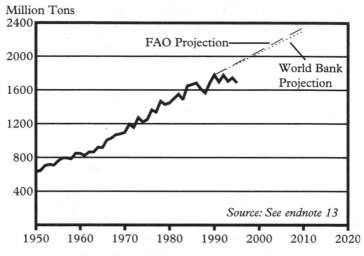

FIGURE 2-6. *World Grain Production, 1950–95, With FAO and World Bank Projections, 1990–2010*

"historically, yields have grown along a linear path from 1960 to 1990, and they are projected to continue along the path of past growth."[14]

To justify this approach, the Bank argues that since the past is the only guide to the future, extrapolation is the most appropriate technique. To project wheat yield per hectare, for example, Bank economists take the wheat yield per hectare in the United States in 1960 and in 1990, and draw a line between the two. This line is then simply extended to obtain the wheat yield for the year 2010.[15]

Although past production trends certainly must be taken into account, the recent past offers evidence that emerging constraints are making it more difficult to expand production rapidly. A well-developed body of scientific literature in the field of production biology describes the trends in biological growth processes in finite environments: they invariably show S-shaped growth curves; none continue to grow indefinitely.

For example, from 1900 until 1950 the world fish catch increased slowly, and then it accelerated after mid-century, climbing steeply in the sixties, seventies, and eighties before levelling off during the nineties, yielding an S-shaped growth curve as the size of the catch was eventually limited to the sustainable yield of fisheries. Whether it is the capacity of the oceans to supply seafood or the rise in grain yield per hectare, any biological growth process in a finite environment eventually slows and levels off.[16]

If the World Bank and FAO want to produce more realistic food production projections, they should replace their teams of economists with interdisciplinary teams that can analyze the new forces that are influencing the food prospect. There is a need for an economist's input, but one that is in balance with inputs from other key fields.

A hydrologist is needed, for instance, to assess the rate of aquifer depletion, estimating when aquifers that are currently overpumped will be depleted and how this will affect irrigation water supplies. With water tables falling in key food-producing regions, any projections that do not take aquifer depletion into account will overstate future food production trends.

Reliable food production projections will also require input from land use experts who can estimate future cropland losses to nonfarm uses, particularly in Asia, which is industrializing at a record pace. This calculation should also include land lost to the development of automobile-centered transportation systems, such as those now planned for India and for China, which wants to expand its car fleet from 2 million in 1995 to 22 million in 2010. In these countries, the automobile is threatening future food security as cropland is paved over for roads, highways, and parking lots.[17]

Also needed on the interdisciplinary team is a plant physiologist—someone who understands the physical limits of plants to absorb nutrients and how this in turn constrains the use of fertilizer to raise yields. With farmers facing diminishing returns on the use of additional fertilizer, the future contribution of this input to growth in the world grain harvest needs to be reassessed.

A meteorologist could estimate the effect of the continuing rise in atmospheric concentrations of carbon dioxide and other greenhouse gases on the frequency and intensity of heat waves. In the past, there was little need to be concerned about climate change in agricultural projections. But with atmospheric carbon dioxide levels now moving to new highs each year, farmers may be soon facing temperatures far higher than any experienced since agriculture began 10,000 years ago.[18]

Without an agronomist to assess and incorporate the effect of soil losses from wind and water erosion on future land productivity, projections will overstate future production gains. Because soil erosion is a gradual process and because data are often lacking on the amount of topsoil lost each year, the effect on food production is often simply ignored.

A marine biologist on the projections team could assess the capacity of oceanic fisheries. If, as now seems likely, the world's farmers do not get any help from fishers in expanding the world food supply, this needs to be taken into account in considering the food supply/demand balance.

Although this is not an exhaustive list of disciplines that can make useful inputs, it does give a sense of some of the fields of knowledge, other than economics, that can help improve the accuracy of projections. Last, there is also a need for an economist who can realistically assess the production response to higher prices under the changing conditions discussed here, rather than simply assuming that it will be the same as in the past.

When grain prices doubled in the mid-seventies, farmers invested heavily in new irrigation wells to boost output. In the late nineties, such an investment may simply accelerate the depletion of aquifers. Similarly, higher grain prices sharply boosted fertilizer use 20 years ago. But in many countries, investing in additional fertilizer now will have little effect on production. And extensive additional investment in fishing trawlers boosted the world fish catch two decades ago. Such additional investment in the late nineties, however, will simply hasten the collapse of oceanic fisheries. In summary, there will be a positive production response to higher food prices in the late nineties, but it is likely to be muted compared with those in the past.[19]

Until recently, Worldwatch Institute was almost alone in arguing that the FAO and World Bank projections were misleading. But this is beginning to change. At the end of 1995, Japan's Ministry of Agriculture released a set of global agricultural supply and demand projections showing that the future could bring scarcity, not surpluses, and a doubling of grain prices. In projections similar to ours, the Ministry—using 1992 as a base and assuming a more or less business-as-usual approach to agriculture—projected that by 2010 the price of wheat would increase 2.12 times and that of rice by 2.05 times.[20]

The difference between the projections by the World Bank and FAO on the one hand and Worldwatch and the

Japanese government on the other is not merely one of degree, but of direction: The former say that world grain prices will continue their long-term historical decline through at least 2010. The latter argue that they will rise, more than doubling by 2010.

If the assessments of the Institute and the Japanese government are reasonable, the future will not be a simple extrapolation of the past. This means that the World Bank and FAO are overestimating food production and, hence, misleading political leaders. This, in turn, may be leading to costly underinvestment in agriculture and family planning.

If the future projected by the Institute and the Japanese government materializes, then the world is faced with steep rises in food prices. In Third World cities, where hundreds of millions of people will be trapped between subsistence-level incomes and rising food prices, hunger will spread. What is not yet clear is how this projected scarcity will affect not only economic stability, as countries wrestle with the inflationary effects of rising food prices, but also political stability. In a global economy that is more integrated than ever before, political instability in any major country or region can affect economic trends everywhere.

I

The Growing Imbalance

3

Demand for Grain
Soaring

Growth in the world demand for grain is being driven
both by the increase in world population and by the rise
in affluence that is enabling so many of the world's
lower-income consumers to move up the food chain, eating more grain-intensive livestock products. The growth
in human numbers and the rise in incomes are both proceeding at record or near record rates.

As noted earlier, those of us born before 1950 have
witnessed a doubling of world population during our
lifetimes. After peaking at roughly 2.2 percent in 1962,
the annual growth in world population declined to less
than 1.6 percent by 1995. Nonetheless, because the
population base is still expanding, the number of people
added each year climbed from just over 70 million in
1962 to nearly 90 million in 1995. (See Figure 3-1.) In

Million

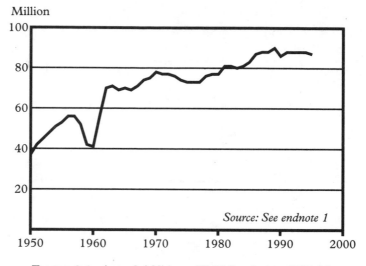

FIGURE 3-1. *Annual Addition to World Population, 1950–95*

other words, although the rate has slowed, it has not slowed fast enough to offset the growth in the population base.[1]

The growth in world population is not at all evenly distributed. Of the nearly 90 million people added each year, more than 80 million are added in developing countries. Asia alone accounts for 54 million of these, some 60 percent of the global total.[2]

The rise in affluence during the nineties is also concentrated in the developing world, especially in Asia. As noted in Chapter 1, in each of the last three years the Asian economy—Pakistan east through Japan, but excluding Japan—has grown more than 8 percent a year. There is no precedent for either the scale or the rapidity of the rise in affluence occurring today in Asia. China, the largest country in the region—with 1.2 billion people today and projected to top 1.6 billion before stabilizing—is also the world's fastest growing economy in the nineties.[3]

Economic growth rates in Asia during the nineties are much higher than had been projected. In 1995, the Chinese economy expanded 10.2 percent (its slowest rate in four years). South Korea was close behind, at just under 10 percent. Vietnam hit 9 percent and Thailand, 8 percent. India, a country of 931 million people, is also beginning to pick up the economic pace, reaching nearly 6 percent in 1995. Indonesia, with a population of 198 million, is expanding at nearly 8 percent a year. Among the economies of any size in the region, Bangladesh was the slowest growing, at 4.9 percent. Barring some unforeseen development, these economic growth rates are projected to continue at close to this pace in the years immediately ahead.[4]

The way that population growth raises the demand for grain is rather simple and straightforward: providing for the growth in population requires some 28 million additional tons of grain each year, or 78,000 tons a day. The influence of rising affluence is much more complicated. Those who live in low-income countries, such as India, consume roughly 200 kilograms of grain a year, just over 1 pound per day. Nearly all this small amount of grain must be eaten directly to satisfy minimal caloric needs; little can be converted into meat, milk, eggs, or other livestock products. In affluent societies, such as the United States or Canada, grain use per person is roughly 800 kilograms a year, most of which is consumed indirectly in the form of livestock products. (See Table 3-1.)[5]

In India, the per capita consumption of eggs, a popular form of protein in a society where religious restrictions limit meat consumption, totals roughly 30 eggs per year, or one egg every two weeks. (The average

TABLE 3-1. *Annual Per Capita Grain Use and Consumption of Livestock Products in Selected Countries, 1995*

Country	Grain Use[1]	Beef	Pork	Poultry	Mutton	Milk[2]	Eggs[3]
				(kilograms)			(number)
United States	800	45	31	46	1	288	174
Italy	400	26	33	19	2	197	158
China	300	4	30	6	2	5	196
India	200	1	0.4[1]	1	1	34	<30

[1]Rounded to nearest 100 kilograms. [2]Estimates based on FAO production figures for 1994. [3]Total number of eggs per person.
SOURCE: See endnote 5.

American ate 174 eggs in 1995.) In all other products, however, the difference between India and industrial countries is even wider. Annual consumption per person of milk, including the milk equivalent of milk products, in India is just over 34 kilograms, compared with 288 kilograms in the United States and 180 kilograms in Italy. For Americans, this means not only drinking a lot of milk directly, but also consuming large amounts of cheese, yogurt, and ice cream.[6]

With meat, the contrast in consumption per person is even greater—from 3 kilograms a year in India, to 42 kilograms in China, to 70 kilograms in Italy and 123 kilograms in the United States. The Chinese eat as much pork and eggs per person as people living in the industrial West, but consumption of beef, poultry, and milk is closer to that of India. For India, the consumption of poultry is expanding by some 15 percent a year, making this the dominant source of meat. Americans, by contrast, eat large quantities of beef, pork, and poultry.[7]

World meat production has increased steadily since 1950, rising nearly every year during this 45-year span. Except in 1960 and again in 1973, a year when world grain prices doubled, it has been one of the world's most predictable economic trends. Turning to meat as incomes rise appears to be an innate tendency, perhaps reflecting 4 million years of human evolution when hunting and gathering were both essential to our survival.[8]

Between 1950 and 1995, world meat consumption climbed from 44 million to 192 million tons, an increase of more than fourfold. (See Figure 3-2.) The 1995 increase of 8 million tons, much of it concentrated in China, was the largest on record. The desire to eat more meat is both strong and pervasive, so much so that it continues to boost meat intake even during economic downturns. In per capita terms, meat consumption worldwide climbed from 17 kilograms in 1950 to 32

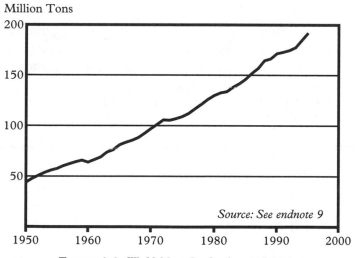

FIGURE 3-2. *World Meat Production, 1950–95*

kilograms in 1988, nearly doubling. After hovering at that level through 1994, the world average topped 33 kilograms in 1995, boosted largely by the rise in affluence in Asia.[9]

The world's population today divides roughly into three categories in terms of position on the food chain. For nearly a billion of the world's people, the most affluent one fifth of humanity, diets are largely saturated with livestock products. In North America and most of Europe (including the European portion of the former Soviet Union), Japan, Australia, and Argentina, there is little desire to move further up the food chain. Indeed, in Western Europe total grain consumption has actually declined slightly over the last decade, reflecting both a levelling off in consumption of livestock products and a somewhat more efficient conversion of grain into those products. A few countries actually moved down the food chain in recent years; in Germany, for example, both beef and pork consumption were somewhat lower in 1995 than in 1990.[10]

Another group, totalling close to a billion people, consists of those who live in poor countries, where incomes are not rising. They are not diversifying diets by moving up the food chain, though they would very much like to. Much of Africa is in this category.[11]

The remaining 3.7 billion people live in low- or middle-income countries where incomes are rising steadily, enabling them to follow the path of the affluent. This includes, for example, most of Asia and the Middle East, most countries in Latin America, and a few in Africa. As noted earlier, this movement up the food chain in the nineties is concentrated in Asia. In China, for instance, pork consumption climbed from less than 23 million tons in 1990 to an estimated 37 million tons in 1995, a phenomenal gain of some 60 percent in just five years. Today, China accounts for half of world pork consumption. Its use of beef during the same period quadrupled, from 1.1 million tons to an estimated 4.4 million tons.[12]

As economic expansion accelerates in the Indian subcontinent, use of livestock products is rising there as well. Both the poultry and dairy industries are growing rapidly. Although starting from very low levels, the production of livestock products, including poultry, milk, eggs, and even beef, is climbing. India's broiler industry, for example, had roughly 30 million chickens in 1980, but reached 300 million by 1995. Growing by some 15 percent annually, its output is doubling every five years. Milk production has expanded by 4 percent in each of the last two years. Production of beef and veal in 1995 totalled 1.1 million tons, up 12 percent from the year before.[13]

Although Indonesia is a small country compared with China and India, it is large by standards outside Asia. With an economic growth rate in the upper single-digit range boosting incomes in recent years, the consumption of poultry and eggs is rising rapidly there, a trend that is expected to continue for the foreseeable future. The demand for meat is also escalating in several smaller Asian countries, such as Thailand, Malaysia, and the Philippines, reflecting their rapid economic gains.[14]

Worldwide, the grain used to feed livestock, poultry, and fish in 1995 totalled 640 million tons, accounting for 37 percent of global grain consumption. During the preceding 30 years, feedgrain use ranged consistently between 37 and 40 percent.[15]

The largest user of feedgrains, not surprisingly, is the United States, which now feeds roughly 160 million tons to cattle, pigs, poultry, and fish each year. Other major feed users are the European Union and China. After showing little growth from 1950 to 1978, China's feedgrain use has increased fivefold since the economic reforms that year. Between 1994 and 1995, China's use of grain for feed jumped from 82 million to 95 million tons even though the domestic price of corn, its dominant feedgrain, doubled between early 1994 and late 1995, rising well above world market levels. (See Figure 3-3.) In 1996, usage is expected to literally go off the top of the chart. Four consecutive years of double-digit economic growth, with a fifth in prospect for 1996, are driving the demand for livestock products up at a feverish pace.[16]

While feedgrain use has been soaring in China since 1990, it has been plummeting in the former Soviet Union. One consequence of economic reforms launched there in 1988 and the union's breakup into its

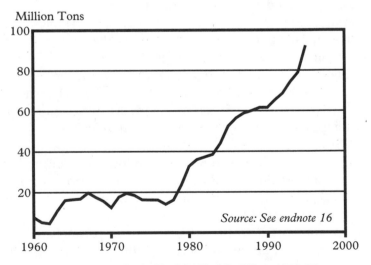

FIGURE 3-3. *Grain Used for Feed in China, 1960–95*

constituent republics was a dramatic reduction in incomes and in consumption of livestock products. Grain use for feed in the former Soviet Union dropped from 149 million tons in 1990 to 75 million tons in 1995.[17]

One result of the unprecedented growth in the demand for livestock products by Asia's increasingly affluent consumers is an extraordinary growth in the feed business. Mills that grind grain and incorporate protein supplements, such as soybean meal, into nutritionally balanced mixed-feed rations are a major growth industry. New mills are now being started almost every day. Literally hundreds of new feed mills are being built each year, most of them in fast-growing economies such as China, India, Indonesia, Malaysia, South Korea, and Thailand. China is the clear leader, but the market for mixed feed is expanding throughout Asia. If the regional economy continues to prosper in the years ahead, thousands of additional feed mills will be needed.[18]

Livestock, poultry, and fish vary widely in the efficiency with which they convert grain into animal protein. Beef cattle in the feedlot typically require roughly 7 kilograms of grain for each kilogram of body weight added. Grass-fed livestock, of course, require no grain at all. Pigs, which cannot digest large amounts of roughage, rely heavily on grain or other concentrated feeds. Typically they require 4 kilograms of grain for each kilogram of additional pork. For cheese and eggs, the ratio is roughly 3 to 1. Poultry and fish are the most efficient at converting grain into protein. In modern broiler operations, roughly 2.2 kilograms of grain are required for each kilogram of chicken. For fish, such as catfish or carp, less than 2 kilograms of grain is needed per kilogram of live weight.[19]

These contrasting efficiencies are beginning to alter patterns of world meat consumption. (See Figure 3-4.) Since the economic reforms in 1978, China's rise in affluence has sharply boosted pork production, largely because the country lacks the rangeland to produce large quantities of beef. Now that the world's rangelands are being fully used to produce beef and mutton, additional beef can come only from feedlot operations, but given the inefficiency of the conversion, beef is losing out to pork and, even more, to poultry. Within the next two years, world poultry production is likely to overtake beef for the first time in history.[20]

Asia is also rapidly raising its intake of beer, another activity that puts pressure on grain supplies. This trend in China, which now has some 800 breweries, appears to be closely tracking the earlier growth in Japan, where beer consumption per person is now among the highest

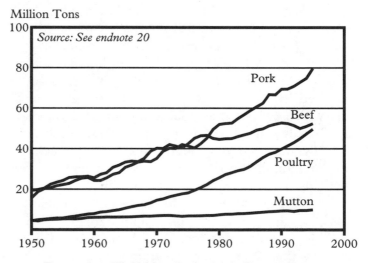

FIGURE 3-4. *World Meat Production by Type, 1950–95*

in the world. Already China has overtaken Germany in total beer consumption, leaving it second only to the United States.[21]

In some regions, growth in the demand for grain is almost exclusively a result of population growth. In others, rising affluence is the source. In still others, it is a combination of the two. And in Western Europe, the use of grain is no longer growing, now that population growth has stabilized and there is no further movement up the food chain.

The main region where population growth is responsible for the additional consumption of grain is Africa. In the United States, where there is little movement up the food chain, it is also mostly population growth. In Asia and Latin America, both are at work.

In some countries where incomes are rising steeply, such as China, movement up the food chain accounts for most of the growth in the demand for grain. For example, between 1990 and 1995, grain consumption in China climbed from 328 million to 368 million tons. Of this 40-million-ton gain, 33 million tons—more than four fifths—went for feed. To put this in perspective, this five-year growth in China's use of grain for feed easily exceeds the 1995 Australian grain harvest of 26 million tons.[22]

The ability of the world's farmers to expand grain production in the years ahead would be challenged by either the growth in population or the rise in affluence. Together, these trends promise a record growth in the demand for grain, posing a formidable challenge for the world's farmers and an even more formidable one for its political leaders.

4

Land Hunger
Intensifying

Ever since agriculture began some 10,000 years ago, farmers have been working to expand cultivated area. Initially, they moved from valley to valley and then later from continent to continent, always looking for fertile new land to plow.

As population grew, ingenious methods were devised for expanding the area used to produce crops. These included irrigation, terracing, drainage, fallowing, and even, for the Dutch, reclaiming land from the sea. Terracing let farmers cultivate steeply sloping land on a sustainable basis, quite literally enabling them to farm the mountains as well as the plains. Drainage of wetlands opened fertile bottomlands for cultivation. Alternate-year fallowing to accumulate moisture helped farmers extend cropping into semiarid regions. But by mid-century, the frontiers of agricultural settlement had largely disappeared, leading to a dramatic slowdown in expansion of the area in crops.[1]

Reviewing cropland trends in the recent past helps to understand the long-term food prospect. Since 1950, growth in the grain area has come in two surges. (See Figure 4-1.) The first occurred in the fifties, when the Soviets opened up the so-called virgin lands to cropping. Concentrated in Kazakstan, this led to the plowing of 25.5 million hectares of virgin grasslands and the planting of spring wheat on an area of newly plowed land roughly equal to the wheat area of Canada and Australia combined.[2]

After grain prices doubled in the mid-seventies, there was a second big push to expand cultivated area, this time in the United States as well as the Soviet Union. Both the virgin lands expansion during the fifties and the

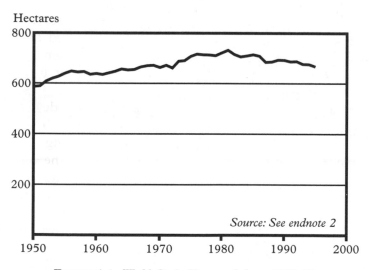

FIGURE 4-1. *World Grain Harvested Area, 1950–95*

push in the mid-seventies involved plowing land that was highly erodible and could not sustain cultivation over the long term.

The historic peak in world grain harvested area came in 1981 as the U.S. government lifted all acreage controls, permitting farmers to plant from fencerow to fencerow. In the former Soviet Union, meanwhile, grain area had peaked and begun to decline as farmers abandoned some of the highly erodible land that had been plowed. The peak 123 million hectares harvested there in 1977 had dropped to 91 million hectares by 1995, declining almost every year during this 18-year stretch.[3]

Much of the more recent cropland loss was in Kazakstan. After reaching 25 million hectares in the mid-eighties, the area sown to grain there began to decline, shrinking to only 18.6 million hectares in 1995 as marginal land was abandoned. A study by the Kazakstan Agricultural Academy of Science sees this shrinkage continuing before levelling off at 16.3 million hectares. A second analysis, this one by the country's Institute of Soil Management, indicates that grain production is sustainable on only 13 million hectares, suggesting an eventual additional loss of more than 5 million hectares of cropland. Whichever turns out to be closest to the mark, further losses lie ahead. The progressive effect of soil erosion, mostly from wind, and other forms of land degradation in this region of low rainfall (less than 300 millimeters a year) has reduced yields on the marginal land areas to less than 500 kilograms per hectare, well below that in most African countries.[4]

In the United States, a more formal effort was made to rescue the highly erodible land that was plowed in response to the high prices of the mid-seventies. In 1985, Congress, with the strong support of environmental groups, passed the Conservation Reserve Program (CRP), an initiative designed to retire much of this land by paying farmers to return it to grass before it became wasteland. So in both the former Soviet Union and the United States the expansion of cultivated area in response to higher prices led to a major retrenchment.[5]

After land degradation, another leading source of cropland loss is industrialization, a loss that is particularly pronounced in countries that are densely populated before rapid industrialization gets under way. The changes during industrialization claim large amounts of land for the construction of factories and warehouses, as does the evolution of an automobile-centered transportation system. But the associated rise in incomes also generates a demand for agricultural products other than grain, such as vegetable oils, fruits, and vegetables.

All these factors have been at work in Japan, South Korea, and Taiwan. Japan's grainland area peaked in 1955 and has shrunk by 52 percent since then. South Korea's peaked in 1965, a decade later, and has dropped 46 percent. For Taiwan, the peak year was 1962, and the decline since then, 42 percent. The combined grainland area in these three countries has shrunk by 48 percent from the peak, or by more than 1 percent a year.[6]

Part of the grainland that was lost was used for industrial and residential construction and for the highways, roads, and parking lots needed for automobiles. Another part of it was shifted to the production of fruits and vegetables and other high-value crops to satisfy the diversification of diet associated with rising affluence.

As Asia industrializes, the construction of thousands of factories, roads, parking lots, and new cities is eating into its remaining productive cropland. In China, the world's most populous country, farmland is being claimed not only by factories, housing, and roads but also by shopping centers, tennis courts, golf courses, and private villas. In rapidly industrializing Guangdong Province, an estimated 40 golf courses have been built in the newly affluent Pearl River delta region alone. In 1995, concern about the effect on food production of this wholesale loss of cropland to recreational uses led the Guangdong Land Bureau to cancel the construction of all golf courses planned but not yet completed.[7]

China has experienced a particularly rapid loss of cropland in the southern coastal provinces, where much of the rice crop is produced. A combination of rapid industrialization and conversion of cropland to other nonfarm uses has taken such a heavy toll that it led to an actual decline in rice production from 1990 to 1995. In simple terms, the loss of riceland more than offset the rise in land productivity, reducing the harvest.[8]

Other countries are now facing heavy losses. Each year, rapidly industrializing Indonesia is losing an estimated 20,000 hectares of cropland on Java alone, enough to supply rice to 280,000 people. In Vietnam, Prime Minister Vo Van Kiet announced a ban on building factories in rice paddies in the spring of 1995 as he tried to preserve the country's cropland. Just four months later he changed his mind—in order to allow Ford Motor Company and other firms to build on 6,310 hectares of farmland near Hanoi.[9]

This was not the only exception. When Daewoo Development, a South Korean conglomerate, protested that the ban would interfere with the completion of the Van Tri Marsh golf complex containing hotels, offices, and apartments, its request for an exemption was granted. Daewoo had argued that Vietnam's future lay with industrialization, not with agriculture. Prior to the ban on riceland conversion, Vietnam was losing an estimated 20,000 hectares of riceland annually to other uses.[10]

To make matters worse, industrialization is taking some of the region's best cropland. For example, land now occupied by factories in southern China was just a few years ago producing two or three crops of rice per year. This is some of the most productive cropland not only in China, but in the world.

A similar loss is occurring in California. In the state's Central Valley, with the world's largest concentration of fruit and vegetable production, highly fertile cropland was being parcelled out for residential and commercial construction at the rate of 17,000 hectares annually in the early nineties. The American Farmland Trust, interested in protecting the cropland that yields a $13-billion annual harvest of produce in the valley, calculates that development zoning requiring six houses per acre instead of three would save 200,000 hectares from development by 2040.[11]

Another factor affecting the grainland area is the diversion of irrigation water to nonfarm uses. In some parts of the world, such as Arizona in the United States or Saudi Arabia, when aquifers are depleted or water is diverted to urban uses, natural rainfall is so low that the land simply reverts to desert. When subsidies were withdrawn in Saudi Arabia for pumping from a fossil aquifer several hundred feet below the earth's surface, the area in wheat fell precipitously, dropping from 800,000 hectares in 1993 to 470,000 hectares in 1995.[12]

At least one tenth of the world's irrigated area is now afflicted with waterlogging and salinity, a threat almost as old as irrigation itself. Indeed, historians attribute the decline of some early civilizations in the Middle East to the waterlogging and salinity of their irrigation systems.[13]

When surface water from rivers is diverted for irrigation, some of the water evaporates into the atmosphere,
some is used by the plants, and some percolates downward. In some cases, the downward percolation of water
raises the underground water table. (This is in contrast
to the situation where the water table falls as an underground aquifer is depleted.) Over time, the water table,
which may be easily 100 feet below the surface when irrigation with surface water begins, slowly rises. When it
gets within a few feet of the surface, the waterlogging of
the soil impairs the development of deep-rooted crops,
lowering yields.

When the water table is within several inches of the
surface, the water begins to evaporate through the
remaining soil into the atmosphere. As it does, the minerals and salts that all water contain are left on the surface. At some point, the soil becomes too salty for crops
to grow, and the land is abandoned.

The most common way of reversing the process is to
drill wells and pump out some of the excess drainage
water, using it for irrigation. Over time, this lowers the
water table and begins to redistribute the salts downward. Although this is a costly, time-consuming process,
it is often the only way to reclaim land rendered infertile
by waterlogging and salinity.

One of the leading potential cropland claimants in
Asia, particularly in China and India, is the automobile.
In 1995, China had only 2 million cars—barely 1 percent of the U.S. fleet. An increase to 22 million cars by
2010, as now projected, would lead to land being paved
for a national network of highways and roads and for
streets, parking lots, and service stations on a scale that
will inevitably take a severe toll of the country's scarce
cropland. (See Figure 4-2.)[14]

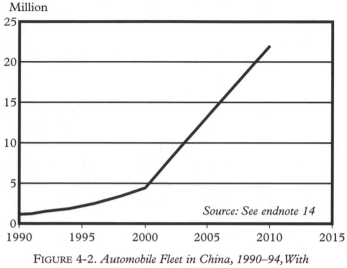

Million

FIGURE 4-2. *Automobile Fleet in China, 1990–94, With Projections to 2010*

India, too, will be sacrificing cropland to the automobile, trading the prestige of car ownership for a few for food security for the many. In fiscal 1996, the Indian automobile industry expanded by an estimated 26 percent. The world's major manufacturers are flocking to the area. Korea's Hyundai plans to invest an additional $1.1 billion, pushing its output in India to 200,000 vehicles annually by 2002. Ford plans to invest $800 millon in two manufacturing facilities over the next seven years. In contrast to China, where debate on similar developments has recently begun, India appears to be giving little serious thought to the automobile's threat to future food security as its population approaches the 1 billion mark in 1999, heading for 1.4 billion by 2030.[15]

The assumption that densely populated developing countries would automatically adopt the automobile-centered transportation system of the industrial world as they industrialize is now being challenged in China. The country's scientific community issued a report that says: "It would be inappropriate for China to encourage the use of family cars on a large scale in the next few years because of the country's serious shortage of land, oil and other resources, and its huge population." Instead, it recommended a "public transportation network that is convenient, complete, and radiating in all directions."[16]

Interestingly, in arguing for this policy shift, the report listed the shortage of land before that of oil, even though China is already importing oil to meet fuel needs. A debate similar to this one between the scientific community and the Ministry of Machine Building Industry, which supports expansion of the automobile industry, is likely to emerge in many other countries as policymakers recognize the trade-off between the growth in private automobile ownership and food security.

In addition to the conversion of grainland to nonfarm uses, substantial areas are converted to other crops, such as oilseeds, fruits, and vegetables, as industrialization progresses and incomes rise. From 1950 to 1995, the world's soybean harvest climbed from 17 million to 123 million tons, a sevenfold gain, as the demand for vegetable oil for cooking and for protein meal supplements for livestock and poultry feed soared. With farmers unable to raise yields per hectare rapidly, as they had with cereals, most of the gains in output had to come from expanding area. As a result, the land in soybeans, which accounts for half of the world oilseed harvest, increased from 14 million hectares in 1950 to 62 million hectares in 1995, with much of the growth coming at the expense of grain. (See Figure 4-3.) In the United States, for example, the cropland in soybeans now rivals that in wheat and corn. The growth in the production and area

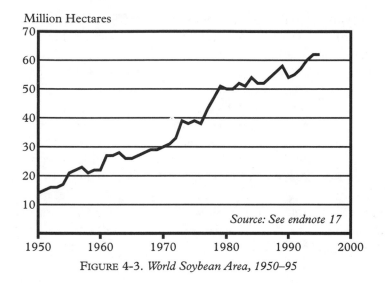

Million Hectares

Source: See endnote 17

FIGURE 4-3. *World Soybean Area, 1950–95*

of soybeans and other oilseeds is a major source of grainland loss since 1981.[17]

As noted earlier, the diversification of diet that comes
with rising incomes means consuming more fresh fruit
and vegetables as well as livestock products. This trend
can be seen quite clearly in China, where the area in vegetables has increased from roughly 3 million hectares in
1978, when economic reforms were introduced, to nearly 9 million hectares in 1994. (See Table 4-1.) Similar
trends can be seen in other countries, such as Japan and
Taiwan.[18]

Despite frequent claims about vast opportunities for
expanding the earth's cultivated area, the opportunities
to do so are in fact quite limited under existing prices.
With a doubling of prices, some marginal land, such as
the *cerrado* (dry plain) in eastern Brazil, might be prof-
itably cultivated. But in this particular case, the plowing

TABLE 4-1. *China: Area in Vegetables, 1970-94*

Year	Area
	(million hectares)
1970	2.7
1979	3.2
1981	3.4
1982	3.9
1983	4.1
1984	4.3
1985	4.7
1986	5.3
1987	5.6
1988	6.0
1989	6.3
1990	6.4
1991	6.5
1992	7.0
1993	7.9
1994	8.7

SOURCE: See endnote 18.

of the *cerrado* is unlikely to do little more than help meet local demand. Brazil, now the largest grain importer in the western hemisphere, is facing a population increase of more than 100 million during the next half-century. If it can become self-sufficient in grain, it will be doing well; it is unlikely to have much left over to export to densely populated countries such as China, Indonesia, or Bangladesh.[19]

The world's grain harvested area can be increased

somewhat by returning to production cropland that is idled under commodity set-aside programs designed to control supply. In the United States, the announcement by the U.S. Department of Agriculture that the modest amount of land still idled under such programs, some 7.5 percent of cornland, would be released for production in 1996 means a likely gain of 2 million hectares in the global grain harvested area. And the European Union declared that in 1996 it would return roughly 1.6 million hectares of the grainland set aside under its supply management program. If it decided to bring all its land back, it might be able to increase the harvested grain area by an additional 4 million hectares.[20]

Another source of additional cropland is the Conservation Reserve Program in the United States, where some 14 million hectares of cropland, much of it highly erodible, has been retired under 10-year contracts. Beginning in 1996, the earliest of these contracts will expire. The lion's share of the 14 million hectares is wheatland, most of it highly vulnerable to wind erosion. Perhaps half of this CRP land could be farmed sustainably with the appropriate cultural practices. If so, this would add 7 million hectares of comparatively low-yield land to the harvested cropland area over the next five years as the CRP contracts expire.[21]

Although the world grain harvested area expanded from 1950 until it peaked in 1981 and then began a gradual decline, the grainland area per person has been declining rather steadily since 1950, shrinking from 0.23 hectares to 0.12 hectares in 1995. (See Figure 4-4.) Even if the grainland area can be boosted from 690 million hectares in 1996 to 700 million hectares in 2030, the area per person will be 0.08 hectares, one third less than today.[22]

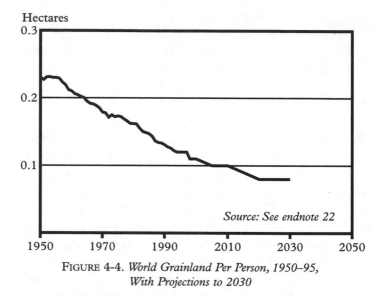

FIGURE 4-4. *World Grainland Per Person, 1950–95,*
With Projections to 2030

Farmers were able to more than offset the shrinkage
in grainland area by raising land productivity over most
of the period since mid-century, but they have not been
able to do this since 1984. Unless they can find a way of
doing so, the world will face a continuing drop in grain
production per person and rising grain prices.

5

Water Scarcity Spreading

The expanding demand for water is pushing beyond the sustainable yield of aquifers in many countries and is draining some of the world's major rivers dry before they reach the sea. As the demand for water for irrigation and for industrial and residential uses continues to expand, the competition between countryside and city for available water supplies intensifies. In some parts of the world, meeting growing urban needs is possible only by diverting water from irrigation.

One of the keys to the near tripling of the world grain harvest from 1950 to 1990 was a 2.5-fold expansion of irrigation, a development that extended agriculture into arid regions with little rainfall, intensified agricultural production in low-rainfall regions, and increased dry-season cropping in monsoonal regions. It also accounts for part of the phenomenal growth in world fertilizer use since mid-century. Most of the world's rice and much of its wheat is produced on irrigated land.[1]

Irrigation is not a new technology. Although its origins are clouded in the mists of history, the available archeological evidence suggests it arose first in the Middle East, roughly midway between the beginning of agriculture some 10,000 years ago and today. The cooperation required for the engineering construction needed to dam rivers and divert the water onto the surrounding land developed hand-in-hand with civilization itself. Only with a sophisticated civil structure was large-scale irrigation possible.[2]

From the beginning of irrigation until 1900, irrigated area slowly expanded, eventually covering some 40 million hectares. From 1900 to 1950, the pace picked up, and the total area more than doubled to 94 million hectares. But the big growth occurred after that. From 1950 to 1993, the irrigated area expanded from 94 million to 248 million hectares. (See Figure 5-1.)[3]

During this period, a key threshold was crossed in 1979. From 1950 until then, irrigation expanded faster than population, increasing the irrigated area per person by nearly one third. This was closely associated with a worldwide rise in grain production per person of nearly one third. But since 1979, the growth in irrigation has fallen behind that of population, shrinking the irrigated area per person by some 7 percent. (See Figure 5-2.) This trend, now well established, will undoubtedly continue and accelerate in the years ahead as the demand for water presses ever more tightly against available supplies.[4]

The 154 million hectares added to the world's irrigated area after mid-century involved the building of thou-

Million Hectares

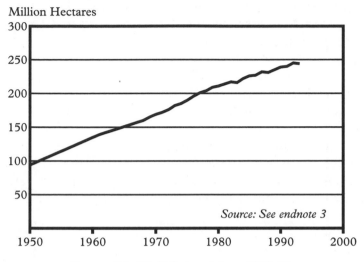

FIGURE 5-1. *World Irrigated Area, 1950–93*

sands of dams to divert river water onto the land and the drilling of millions of wells to tap underground water supplies. Building dams and drilling wells both took large amounts of capital, making irrigation a major focus of public investments by national governments and international development agencies. In addition, farmers drilling wells on their own land made a substantial contribution to these impressive gains.[5]

The best conditions for irrigating with river water are found in Asia, which has some of the world's great rivers—the Indus, the Ganges, the Chang Jiang (Yangtze), the Huang He (Yellow), and the Brahmaputra. These originate at high elevations and travel long distances, providing numerous opportunities for dams and the diversion of water into networks of gravity-fed canals and ditches. As a result, some two thirds of the world's irrigated area is in Asia. China and India lead

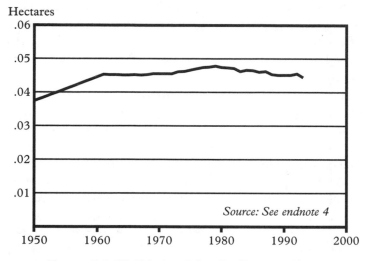

Hectares

FIGURE 5-2. *World Irrigated Area Per Person, 1950–93*

the world, with 50 million and 48 million hectares of irrigated land respectively.[6]

In monsoonal climates, where the wet season is followed by a long dry season, with several months of little or no rain, irrigation often holds the key to cropping intensity. Where temperatures permit year-round cropping, as they often do in such climates, irrigation allows the production of two or even three crops a year. Particularly rapid irrigation growth in China helped increase multiple cropping from an average of 1.3 crops per hectare in 1950 to 1.5 in 1980.[7]

Since the early sixties, the development of underground sources with wells drilled especially for irrigation has become relatively more important than the development of surface water. In India, in particular, the more profitable, high-yielding grain varieties introduced in the mid-sixties led to heavy investments by farmers in their own wells, particularly in the Gangetic Plain. Similarly, much of the irrigation expansion in the United States after mid-century occurred in the southern Great Plains, based on the use of underground water and large center-pivot irrigation systems.[8]

During the nineties, several trends are emerging to reduce the irrigated area. Among these are the depletion of aquifers, the diversion of irrigation water to cities, and the restoration of river flows to protect endangered ecosystems.

As noted in earlier chapters, water tables are now falling in major food-producing regions throughout the world. This is most dramatic where irrigated agriculture depends on fossil water, such as in the southern Great Plains of the United States, Saudi Arabia, and Libya. An estimated 25 percent of U.S. irrigated cropland is watered by the unsustainable practice of drawing down underground aquifers. In Iran, an estimated one third of all cropland depends on irrigation that involves the over-drafting of groundwater.[9]

In the U.S. Great Plains, farmers from South Dakota through Nebraska, Kansas, eastern Colorado, parts of Oklahoma, and the Texas panhandle were able to expand irrigation from mid-century through 1980 by tapping the vast Ogallala aquifer. But this is essentially a fossil aquifer. Although in some locales it does receive a modest recharge from rainfall, most of the water in it was deposited there eons ago. Heavy reliance on it, therefore, is ultimately unsustainable. And in some of its more shallow southern reaches, it is already partly depleted. As a result, between 1982 and 1992 irrigated area in Texas shrank 11 percent, forcing farmers to return to traditional—and less productive—dryland farming. Irrigated area is also shrinking in Oklahoma,

Kansas, and Colorado.[10]

In India, water tables are falling in several states, including the Punjab—the country's breadbasket. The double cropping of winter wheat and rice there has dramatically boosted the overall grain harvest since the mid-sixties, but it has also pushed water use beyond the sustainable yield of the underlying aquifer. Water tables are also falling in the state of Rajasthan's Jodhpur district. As this happens, cities and towns drill deeper wells. Meanwhile, villagers without the capital to deepen their own wells are left high and dry, forced to abandon irrigated agriculture.[11]

In China, which is trying to feed 1.2 billion increasingly affluent consumers, much of the northern part of the country is a water-deficit region, satisfying part of its needs by overpumping aquifers. Under Beijing, for example, the water table has dropped from 5 meters below ground level in 1950 to more than 50 meters. During a visit to Worldwatch Institute, Professor Chen Yiyu, the vice president of the Chinese Academy of Sciences with the responsibility for agriculture and water resources, noted that under a large area of northern China the water table fell some 30 meters over the last two to three decades. He estimates that some 100 million people lived in the affected area.[12]

His knowledge of the number of people likely to be affected suggests that he and his colleagues at the Academy who monitor the depletion of aquifers carefully are very much concerned about the welfare of people in the area where the inevitable reduction in water supplies will come when the aquifer is depleted. Whether that depletion is imminent or is still some years away is not clear. But whenever it comes, it will reduce the amount of water available for this population dramatically. Meeting their water needs for residential and industrial uses may be possible only with a steep cutback in irrigation.

The water situation is particularly acute in Hebei Province, where the depletion of groundwater is shrinking the area of irrigated farmland. As the demand for water in industrial cities within the province has grown, farmers have increasingly been excluded from reservoirs they have traditionally depended on. Aquifer depletion is proceeding so fast that it is leading to subsidence, sinkholes, and fissures. Groundwater depletion around the industrial city of Handan has created some 30 fissures in the land, including one passing through the heart of the city that residents fear may cause buildings to collapse.[13]

In the central and northern provinces of Shanxi, Hebei, Henan, and Shandong, the amount of water available for irrigation has fallen to a fraction of that needed to maximize yields. Vaclav Smil, a China scholar at the University of Manitoba in Canada, observes that the growing water needs of China's cities and industrial areas "will tend to lower even those modest irrigation rates."[14]

Overpumping is a way of converting a minor crisis in the short run into a major crisis in the long run. It lets policymakers avoid difficult questions about carrying capacity and population policy, about the need to create water markets, and about the importance of investing in efficiency. In fact, it is a way to defer tough choices to the next generation, leaving them with even more demanding adjustments.

As the demand for water rises more or less continually with the growth of population, pumping also increases. At some point, the rising use of water crosses the sustainable yield threshold of the aquifer, and the water table begins to fall. Eventually, when the aquifer is depleted and the water supply falls, the rate of pumping necessarily drops to the rate of recharge. If water is being pumped at twice the rate at which an aquifer is being recharged by rainfall, the supply of water will be cut in half when the aquifer is depleted.

In an area like India's Punjab, this may mean that the double cropping of wheat and rice will have to be modified by substituting a lower-yielding dryland crop, such as sorghum or millet, for the rice. For India—adding 16 million people each year—this is not a pleasant prospect.[15]

In northern China, where water tables are falling by 0.5 to 3.0 meters a year in some places and where industrial and residential demands for water are climbing at record rates, heavy irrigation cutbacks are inevitable. Indeed, they are already beginning in some locations. Other areas facing heavy cutbacks include several states in India, among them Punjab, Haryana, Rajasthan, and Tamil Nadu, and several states in the United States, including California and Arizona. And as noted earlier, in the southern Great Plains cutbacks are already under way in Texas, Oklahoma, Kansas, and Colorado.[16]

Perhaps the most extreme case of overpumping a rechargeable aquifer is in Tunisia, where the rate of pumping may be 10 times the rate of recharge. A recently developed hydrological model for Tunisia indicates that within a few years the aquifer will be depleted, forcing a precipitous decline in the rate of pumping, a drop in grain production, and a need for additional grain imports.[17]

The growing demand for water is putting excessive pressure on rivers as well as aquifers. The planet's great rivers are perpetually renewing, but in more populated regions, rivers have been dammed, diverted, and tapped until often there is little water left to continue on its way. In fact, many rivers now run dry before they reach the ocean.

China's great Huang He (Yellow River), which first failed to reach the sea in 1972, now runs dry each year and for progressively longer periods. In the late spring of 1996, it completely disappeared before it reached Shandong Province, the last one it travels through en route to the Yellow Sea. For the farmers of Shandong, who produce one fifth of China's wheat and one seventh of its corn and who depend on the river for half their irrigation water, this is not good news.[18]

At the same time, on the opposite side of the globe, the Colorado River disappears into the Arizona desert, rarely reaching the Gulf of California. In central Asia, the Amu Dar'ya is often drained dry by Turkmen and Uzbek cotton farmers before it reaches the Aral Sea, thus contributing not only to the sea's gradual disappearance but also to the collapse of the 44,000 ton-per-year fishery it once supported.[19]

Draining rivers dry may be rationalized as essential to human food production, but the benefits it confers on one front have to be weighed against the heavy toll it takes on another. Dried-up or diminished outflows threaten the survival of fish that spawn in these rivers. Estuaries that have served as breeding grounds for oceanic species are destroyed. The nutrient flows from the land to the sea that help sustain fisheries are diminished or even disrupted entirely.

Accordingly, some governments are moving to restore river flow to protect these fisheries—even though it means reducing irrigated area. In California, for example, officials have decided to restore nearly a million acre-feet of water to maintaining fish and wildlife habitat in the rivers and streams of the state's Central Valley. In an effort to protect the health of the San Francisco Bay estuary, they have also agreed to limit the amount of fresh water that can be diverted from the rivers feeding into the bay.[20]

California's 1990 population of 30 million is projected to expand to 49 million by 2020, a gain of 63 percent. With urban water demand expected to rise by 54 percent during this period, the amount available for agriculture seems certain to decline, even as the demand for food is climbing.[21]

The dilemma is that as population grows, the resulting increases in urban and industrial demand can be satisfied only by diverting water from the very irrigation needed to supply that population's food. In Colorado, the small town of Thornton, northwest of Denver, has purchased water rights from farmers and ranchers in Weld County on the state's northern border. It plans to build a 100-kilometer pipeline to transport the water as its demands begin to exceed local supplies in the years ahead. Similarly, in 1995 the city of Fukuoka in southern Japan bought rights for irrigation water from some 700 rice growers to avoid a possible water shortage.[22]

In China, officials decided in early 1994 to ban farmers from the reservoirs around Beijing so the water could be used to meet the city's soaring residential and industrial demands. Those requirements are heightened by the population's growing affluence, which increases per capita water use as more people get indoor plumbing.[23]

In the spring of 1996, farmers in Shangdong Province along the Huang He eagerly awaited restoration of the river's flow, only to learn that they would not be given access to the water. All of it, they were told, was needed at the Dongying oil field near the mouth of the river.[24]

With demand in north China fast outstripping the capacity of the Huang He to supply water, extreme scarcity looms ahead. Even now, drinking water is being rationed in some parts of Shanxi Province. The big loser in the eight provinces along the river will be agriculture. As scarcity deepens, priority will go to satisfying the needs of industrial expansion and urban growth.[25]

Within the Middle East, water demands are met largely by three river systems: the Nile, the Tigris-Euphrates, and the Jordan. As of the mid-nineties, nearly all the water from the Nile and the Jordan is being used. Expanding residential and urban demands in these two basins are being met by diverting water from irrigation, which now accounts for 80 percent of the region's water use. Water is already scarce in the region, but family planning is equally scarce. As a result, population growth in the Middle East, which averages nearly 3 percent a year, is steadily diminishing the water availability per person.[26]

The bottom line is that the human demand for water is beginning to press against the limits of the hydrological cycle in many geographic regions. In a landmark 1996 article in *Science* that analyzed the share of the world's freshwater supply appropriated for human use, Sandra Postel, Gretchen Daily, and Paul Ehrlich observe that "fresh water is now scarce in many regions of the world, resulting in severe ecological degradation, limits on agricultural and industrial production, threats to human health, and increased potential for international conflict."[27]

Worldwide, about two thirds of all the water that is diverted from rivers or pumped up from underground is now used for irrigation, so any cutbacks in water supply are likely to affect the food prospect. In regions where all available water is being used, the competition between farmers and cities is intensifying. In this battle, farmers almost always lose. As irrigation water is diverted to urban areas, grain production falls, forcing countries to import more.[28]

Just as the frontiers of agricultural land settlement disappeared a half-century ago, the frontiers of water resource development are disappearing today. And just as the loss of land settlement led to a heavy emphasis on raising land productivity, so the time may now have come to concentrate on raising water productivity.[29]

The world may be about to enter a new era, one in which the overall irrigated area no longer increases. A few irrigation projects are still coming on stream here and there around the world, including one in Turkey that is systematically tapping the remaining unused potential in the Tigris and Euphrates rivers. Vietnam is planning to expand its irrigated area by tapping the waters of the Mekong. But gains from new projects such as these may be offset by losses elsewhere in the world from aquifer depletion and from diversion to cities.[30]

David Seckler, Director General of the International Irrigation Management Institute, believes the losses may now be exceeding the gains, leading to a shrinkage of world irrigated area. If this is happening, as seems likely, then the irrigation water supply per person—which has declined slowly for some years—will decline even faster if population growth continues as projected. Arresting the decline in water availability per person may now depend more on stabilizing population than on anything else that policymakers can do.[31]

6

Rise in Land Productivity Slowing

Since the middle of this century, there has been relatively little new land to bring under the plow. Farmers responded to land scarcity by raising land productivity. Between 1950 and 1995, they raised the grain yield per hectare from 1.06 tons to 2.52 tons, boosting land productivity 2.4 times. Indeed, farmers have raised land productivity more since 1950 than during the 10,000 years since agriculture began.[1]

That's the good news. The bad news is that in recent years the rise in land productivity appears to have slowed. The 1995 yield of 2.52 tons per hectare, depressed by crop-withering heat waves, was essentially unchanged from the 2.54 tons in 1990. (See Figure 6-1.) Is the trend of the last five years an anomaly, meaning that the rapid historic rise in yields will resume? Or

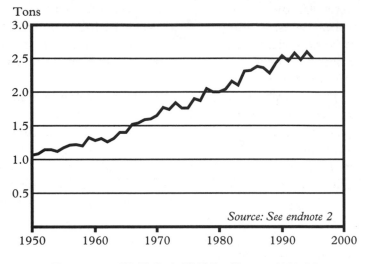

FIGURE 6-1. *World Grain Yield Per Hectare, 1950–95*

is it hinting at a much slower future rise in cropland productivity?[2]

The longer-term global trend of steadily rising land
productivity from 1950 until 1990 masks differing
trends for industrial and developing countries. Much of
the progress in raising grain land productivity during the
fifties and sixties was concentrated in the industrial
countries of North America and Europe, including the
former Soviet Union. By the late sixties, however, yields
were starting to climb in developing countries, particularly in Asia, as the high-yielding wheats and rices were
introduced. During the seventies and early eighties,
gains in land productivity for both wheat and rice in
Asian developing countries were especially impressive.[3]

By 1990, yields in both industrial and developing
countries had reached a level where further gains were
becoming more difficult to achieve. In effect, farmers

were already using most of the available yield-raising technologies. Some countries, such as India, were still increasing yields in a rapid, sustained fashion, but many others had reached a level where it was no longer easy to achieve such results.[4]

The contrast between industrial and developing countries can be seen rather clearly in the yield trends in the world's two largest grain producers: the United States and China, each of which produces one fifth of the world grain crop. In the former, the big gains came in the fifties and sixties, when the yield per hectare rose by roughly 4 percent a year. (See Table 6-1.) After that, it slowed to less than 2 percent in the seventies and to just 1 percent in the eighties. During the nineties, the growth appears to have slowed further.[5]

For China, however, the yield gains came somewhat later and the rate of increase has been highly variable. Yields rose relatively slowly from 1950 until 1977, averaging 2.7 percent a year. (See Table 6-2.) After the eco-

TABLE 6-1. *United States: Change in Grain Yield Per Hectare by Decade, 1950-90*

Year	Annual Yield Per Hectare[1]	Increase by Decade	Annual Gain By Decade
	(tons)	(percent)	(percent)
1950	1.65		
1960	2.40	+ 45	+ 3.8
1970	3.43	+ 43	+ 3.6
1980	4.13	+ 20	+ 1.9
1990	4.56	+ 10	+ 1.0

[1]Each year shown here is actually a three-year average that is used to minimize the effect of weather fluctuations.

SOURCE: See endnote 5.

TABLE 6-2. *China: Change in Grain Yield Per Hectare,*
Selected Years, 1950-94

Year	Annual Yield Per Hectare (tons)	Increase by Period (percent)	Annual Gain By Period (percent)
1950	1.04		
1977	2.11	+ 103	+ 2.7
1984	3.41	+ 62	+ 7.1
1995	4.06	+ 19	+ 1.6

SOURCE: See endnote 6.

nomic reforms of 1978, which broke up the large collective farming operations and returned the responsibility for managing land to individual families, yields increased dramatically. From 1977 to 1984, Chinese farmers achieved a phenomenal annual gain in yield per hectare of 7 percent, much of it due to the soaring use of fertilizer, as they caught up with agriculturally advanced countries. Then from 1984 to 1995, the rate of gain fell precipitously, averaging only 1.6 percent a year.[6]

All the countries that raised yields dramatically from 1950 to 1990 relied on essentially the same combination of higher yielding fertilizer-responsive varieties and a dramatic growth in the use of fertilizer. Some, such as China, India, and the United States, also greatly expanded their irrigated area. As noted in Chapter 1, between 1950 and 1989 the world's farmers increased fertilizer use from 14 million to 146 million tons, a phenomenal tenfold increase. During this period, the growth in world fertilizer use was one of the most predictable of world economic trends.[7]

After 1989, world fertilizer use turned downward,

largely because of the agricultural reforms launched in 1988 in the former Soviet Union. (See Figure 6–2.) The move toward world market prices for fertilizer dramatically boosted the price to farmers, sharply reducing usage. Each year since then, fertilizer use in the former Soviet Union has declined; by 1995, it was less than one fifth of the level in 1989.[8]

In other agriculturally advanced regions, such as North America, Western Europe, and Japan, fertilizer use was levelling off during the eighties. U.S. farmers are actually using less fertilizer during the mid-nineties than they did during the early eighties. Like farmers in other agriculturally advanced countries, they have discovered that at some point applying more fertilizer does not lead to any meaningful rise in grain yields. This realization, combined with the adoption of more careful testing of soils for fertilizer needs, has led to applications that meet crop needs more precisely.[9]

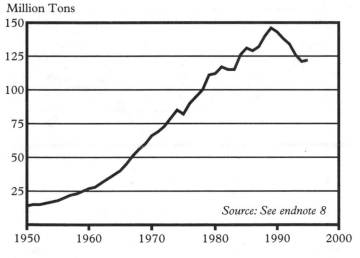

Million Tons

Source: See endnote 8

FIGURE 6-2. *World Fertlizer Use, 1950–95*

Even in some developing countries, the growth in fertilizer use is slowing. For example, in China, which now uses 28 million tons of fertilizer a year (compared with 20 million tons in the United States), the economics of increasing fertilizer use much beyond the current level does not look promising. Indeed, fertilizer use in China may actually decline during the late nineties, much as it did in the United States during the early eighties. (See Figure 6-3.)[10]

In many developing countries, such as India, Bangladesh, and Pakistan, fertilizer use is continuing to rise, albeit slowly. India, for example, is using 14 million tons per year, some 5 million tons less than the United States.[11]

Among the other factors that affect the productivity of the world's croplands are the progressive loss of topsoil from wind and water erosion and higher temperatures,

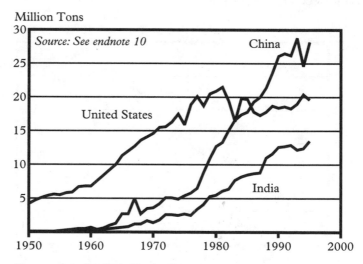

FIGURE 6-3. *Fertilizer Use in China, the United States, and India, 1950–95*

particularly during the summertime in key food-producing regions. Several agronomic studies in the United States on topsoil loss's effect on land productivity report that each inch of topsoil lost reduces yields of both wheat and corn by roughly 6 percent. These productivity losses vary, of course, depending on whether the soil is underlain with subsoil of a lower quality or with rock. If the latter, then the falloff in yield could be much steeper.[12]

Although air pollution is not yet quite as pervasive as soil erosion, it is nonetheless taking a toll on the world's harvests. A detailed study by the U.S. Department of Agriculture and the Environmental Protection Agency, which relied on some 70 monitoring points throughout the United States, determined that air pollution, principally ground-level ozone, was reducing the U.S. crop harvest by at least 5 percent and perhaps by as much as 10 percent.[13]

Studies in Europe to measure the effects of air pollution on crop yields reinforce these findings. Sweden, for example, pays for air pollution with an estimated loss in its grain harvest of 350,000 tons. In the former Czechoslovakia, with some of the world's worst air pollution, the loss of all crops from this source in the mid-eighties was estimated to be equivalent to 1.3 million tons of wheat. If Asian countries, including China and India, continue to expand the burning of coal rapidly as projected, the damage to their harvests from air pollution could be even greater than that associated with fossil-fuel burning in the United States.[14]

Rising temperatures could also lower yields. In 1995, the warmest year on record, crop-withering heat waves in the entire northern tier of industrial countries—the United States, Canada, the European Union, the

Ukraine, and Russia—sharply reduced the world grain harvest. The United States also saw harvests reduced by unusually hot summers in 1980, 1983, 1988, and 1991. In some of these years, intense heat was accompanied by drought. In 1988, severe heat and drought throughout the Midwest dropped grain production below consumption for the first time in U.S. history. Intense heat during the summer growing season not only reduces yields, it leads to much more erratic production trends because it can do so much damage to a crop.[15]

Although there is a great deal of uncertainty about future climate change, the relationship between rising levels of greenhouse gases and the earth's temperature—the so-called greenhouse effect—is well established. Atmospheric carbon dioxide levels have risen every year since measurements began in 1959, climbing roughly 14 percent over the last 36 years. And scientists have established that the average temperature is rising, with the 11 warmest years since recordkeeping began in 1866 all coming since 1979. (See Figure 6-4.)[16]

Given the central role of fossil fuels in the world energy economy, atmospheric carbon dioxide levels are certain to rise much higher in the years ahead. Aside from the likelihood of more intense heat waves, no one knows for sure how global weather patterns will be affected. The world is entering uncharted territory on the climate front, moving toward a climatic regime very different from any since agriculture began.

Assessing the prospect for raising yields is sometimes confused by a lack of understanding of the physical environment associated with high yields. For example, many observers wonder why the Green Revolution has not come to Africa. The reason is the same as the one that kept it out of Australia—it is largely a semiarid and arid

Degrees Celsius

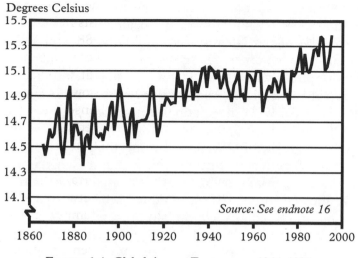

FIGURE 6-4. *Global Average Temperature, 1866–1995*

continent.

In the parts of Africa that have both good soils and adequate water supplies, such as the Nile River valley in Egypt and parts of the East African plateau, or where rainfall levels and seasonal distribution are favorable, including southern Kenya and Zimbabwe, yields have indeed increased rather impressively. But for the continent as a whole, aridity is a major constraint on the long-term food-producing potential. Without water, farmers are seriously limited in the amount of fertilizer they can use effectively. And without fertilizer, they lack the principal input used to raise the productivity of their land.

Latitude also affects crop yield potential. A study by Donald Mitchell and Merlinda Ingco of the World Bank shows that rice yields, which are lowest in the equatorial regions, gradually increase in the higher latitudes. For example, rice yields are much higher in Japan than in

Indonesia. Even if the management practices and inputs used in the two countries are identical, Japan's yields will always be higher because days are much longer there during the summer rice-growing season, leading to greater photosynthetic activity and higher yields.[17]

A similar situation exists for wheat: growing conditions explain wide international variations in yield. In the northern latitude countries that grow wheat, yields are much higher in Western Europe than in Russia or Canada. The harsh winters in Russia and Canada prevent farmers from growing winter wheat, forcing them to sow mostly spring wheat. Winter wheat is typically planted in the fall, usually in late September. It germinates and reaches several inches of height before the onset of winter, when it is dormant. When spring comes and temperatures begin to rise, the wheat is well established and ready to grow rapidly, maximizing growth during the long days of late spring and early summer.[18]

Spring wheat, by contrast, is typically planted in May and harvested in late August or September, which severely limits the length of the growing season and hence the yield potential. In Western Europe, which is warmed by the Gulf Stream and which has rainfall patterns favorable to wheat, yields can easily triple those in Canada or Russia.[19]

Variations in solar intensity can also influence yields. In areas like the Philippines, Indonesia, or southern India, where the temperature permits the double cropping of rice, the dry-season rice crop typically yields 20 or even 30 percent more than the crop grown during the monsoon season. During the dry season, solar intensity is much higher, leading to more photosynthetic activity and higher yields than during the monsoon season, when there is a heavy cloud cover much of the time.

Comparing rice yields in Japan with the much higher rice yields achieved in California and using the latter to calculate Japan's yield potential is misleading simply because solar intensity is so much greater in California.[20]

There is often a tendency both within countries and at the international level to use the highest yields achieved for wheat, rice, or corn as a reference point for calculating the potential elsewhere or for the entire world. This is simply not realistic because of the variations in rainfall, temperature, latitude (day length), solar intensity, and inherent soil productivity just described.

An interesting new technology coming into use in the United States, which is getting a lot of media attention, is commonly referred to as "precision" agriculture. This depends on the use of global positioning satellites (GPS) that can pinpoint the location in a farmer's field within one meter. This enables farmers who are using harvesting equipment with a receiver on it to monitor yield carefully in a highly localized manner within a large field. A four-wheeled vehicle, also with a receiver, is used to sample soils, again for localized areas within the field. A comparison of soil nutrient content, organic matter content, and soil moisture with yields enables a farmer to apply needed inputs, such as fertilizer, more precisely. In some parts of the field, the comparison of yield and soil tests might suggest the need for more fertilizer; in others, less.[21]

The GPS is best suited for areas with rolling topography, like that found in some midwestern states, such as Iowa or Indiana. Its purpose is not so much to raise yields, though it may contribute to a modest increase of a few percent, as it is to boost the efficiency with which inputs are used, such as fertilizer, herbicides, and insecticides. Because of the large-scale costly equipment needed to use this technology, it will likely be confined in the foreseeable future to the United States, Canada, and Western Europe.[22]

One of the keys to assessing the potential for reestablishing a rapid rise in land productivity is whether there are any new technologies in the research pipeline that can lead to quantum jumps in the world grain harvest, such as those associated with the adoption of hybrid corn or the use of fertilizer. There do not seem to be any.

The technologies that have been at the center of record rises in land productivity since mid-century were all developed between 1840 and 1930. For example, the discovery that all the nutrients removed from the soil could be replaced in mineral form was made by Justus von Leibig, a German agricultural chemist, in 1847. The basic principles of genetics that are used in plant breeding were developed by Gregor Mendel, the Austrian monk who did his landmark research in the 1860s. The genes that were incorporated in the high-yielding dwarf wheats and rices that were at the heart of the Green Revolution beginning in Asia in the mid-sixties were first isolated by Japanese scientists and incorporated into indigenous wheat and rice varieties during the 1880s. The hybridization of corn that has contributed to a four-fold rise in corn yield per hectare in the United States was commercialized before 1930. The development of irrigation, of course, goes back several thousand years.[23]

The big gains in land productivity since 1950 have come from the global dissemination of these technologies and their refinement and adaptation to local conditions. There has not been a technological advance since mid-century that would lead to a quantum jump in world food production similar to that associated with any of the basic technological advances outlined above.

And this includes biotechnology. One question frequently asked is, What is the potential of biotechnology to create varieties that will yield much more than those developed by plant breeders using more conventional techniques? The answer, of course, is that no one knows with any certainty. Although biotechnology has been around for more than 20 years, it has yet to produce a single commercially successful, high-yielding variety of wheat, corn, or rice, simply because conventional plant breeders have already done whatever they could think of to raise yields meaningfully. The principal contribution of biotechnology to crop production is likely to come in developing strains that are resistant to various diseases and insects. This in turn reduces the need to use pesticides. And to the extent that this brings crop losses below those achieved with pesticides, it will boost yields.[24]

Biotechnology permits plant breeders to do one thing they cannot do with traditional plant breeding—transfer germplasm from one species to another. Beyond this, it is simply another tool in agricultural scientists' tool kit, one that may permit them to do some things faster and others at less cost. But it is not a magic wand that can be waved at food scarcity to make it go away.

Ironically, the one large potential gain in land productivity that is on the horizon comes from rice breeders at the International Rice Research Institute in the Philippines, who are relying exclusively on traditional plant breeding techniques. They are quite literally redesigning the rice plant in order to raise its productivity closer to that of wheat and corn, the two other major grains. Early results indicate that when the new rice strain becomes available, perhaps shortly after the turn of the century, it has the potential to raise yields by 20 percent. If it does this, 70 million tons would be added to the world grain harvest—enough to cover world population growth for 30 months.[25]

Many trends are making it more difficult to raise yields, but one stands out—the diminishing yield response to the use of additional fertilizer. This is the principal reason that land productivity is rising much more slowly in the nineties. Although the grain area per person was shrinking steadily from 1950 to 1990, the world's farmers were able to more than offset the shrinkage in grain area with rises in land productivity. But since 1990, they have not been able to do so.

The growth in fertilizer use from 1950 through 1989 was the engine driving the increase in the world grain harvest. But the engine is losing steam. (See Figure 6-5.) The question of how to expand production, now that additional fertilizer has little effect on yields in many

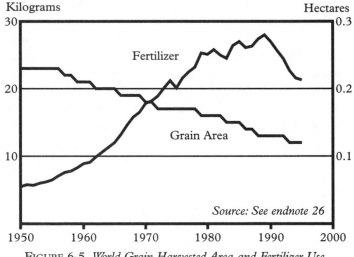

FIGURE 6-5. *World Grain Harvested Area and Fertilizer Use Per Person, 1950–95*

countries, is not just a challenge to farmers. It is an even more serious matter for the world's consumers, whose ranks are expanding by nearly 90 million a year. Figure 6-5 may tell us more about what the future holds than some books written on the subject.[26]

For the world's political leaders, this diminishing response poses a challenge that may be far more difficult than any faced by their predecessors. Whether they can recognize this challenge soon enough to avoid a severe deterioration in the world food situation and a politically destabilizing rise in food prices remains to be seen.

II

Making Choices

7

The New Politics of Scarcity

The signs of a new era in the world food economy are unmistakable. The old formula that was so spectacularly successful in expanding food production for nearly a half-century—combining more and more fertilizer with higher yielding varieties—is no longer working very well. And there is no new formula to take its place. Recent trends in food produced from oceanic fisheries, and croplands give some sense of what the future holds. (See Table 7-1.)[1]

The issue is not whether grain production can be expanded, but whether it can be expanded fast enough to keep up with the record growth in demand. There are many opportunities still for increasing output, including substantial opportunities for productively using more fertilizer in countries such as India and Argentina. But

TABLE 7-1. *Indicators of Food Security in*
Old and New Eras

Indicator	Old Era (roughly 1950 to 1990)	New Era (roughly 1990 into indefinite future)
Grain production per person	Rising: up 40 percent from 1950 to 1984	Falling: down 15 percent from 1984 to 1995
Seafood catch per person	Rising: doubled from 1950 to 1989	Falling: down 7 percent 1989 to 1995; will fall as long as population growth continues
Grain prices	Declining in real terms from 1950 through 1993	Rising: will fluctuate, but around rising trend 1993 onward
Grain stocks	Abundant, often excessive	Low, often inadequate
Idled cropland	Cropland idled throughout this period	Little or no cropland idled after mid-nineties
Grainland per person	Shrinking slowly until 1981, then more rapidly	Shrinking rapidly as long as population growth continues
Irrigated area per person	Expanding: up 28 percent 1950 to 1979	Shrinking since 1979: will continue as long as population growth continues
Fertilizer use per person	Rising: up fivefold 1950 to 1989	Shrinking since 1989: will not rise much as both grainland and irrigation water per person shrink

Effect of climate change	Effect beginning to show as temperatures rise after 1979	More intense heat waves likely to plague efforts to expand output
Backlog of unused technologies	Huge at beginning of era, but diminishing over time	Greatly diminished: no dramatic advance in prospect
Politics of water	Gradually intensifying as period progressed	Intense competition among countries and between countryside and city
Politics of food	Dominated by surpluses; competition among exporters for access to markets	Dominated by scarcity; competition among importers for access to supplies

SOURCE: See endnote 1.

restoring rapid, sustained growth in the grain harvest to the nearly 3 percent a year that prevailed from 1950 to 1990 will be difficult indeed.

In addition to the constraints detailed in Part I, farmers now face the uncertainties of climate change. If heat waves become progressively more intense, as climate models suggest, meeting future food needs could become even more difficult. Crop-withering heat waves of the sort experienced in 1995 in major food-producing regions, such as the U.S. Corn Belt, Europe, the Ukraine, Russia, and Argentina, could become routine. No one knows what the future will bring, but the rise in the earth's temperature since 1979 suggests that the failure to check the increase in atmospheric levels of greenhouse gases could bring temperatures far higher than any since agriculture began.[2]

On the demand side, the growth prospects are much

more bullish, as noted in Chapter 3. Not only is the world continuing to add nearly 90 million people a year, but the unprecedented rise in incomes in Asia, where over half of humanity lives, is fueling a record growth in demand. Just satisfying the growth in world population requires 28 million additional tons of grain per year, 78,000 tons per day.[3]

As the growth in the oceanic fish catch halted and as grain production slowed, the nineties became the first decade in which both seafood and grain production per person were declining. The principal manifestations of this growing imbalance were the fall in world carryover stocks of grain and the return to use of idled cropland under commodity control programs.[4]

During the first half of the decade, the production shortfall was offset by reducing stocks. But with carryover stocks in 1996 at 48 days of consumption—little more than pipeline supplies—the opportunity for further reductions is limited. Stock depletion may mark the transition from the old era of surpluses to the new era of scarcity. Future excesses of demand over supply can be eliminated only by rising prices, which is exactly what happened in late 1995 and 1996. The rise in grain prices that started in China in 1994 soon spread to the rest of the world, helping to double wheat and corn prices by the spring of 1996. Although these prices will decline somewhat in the immediate future, over the longer term higher food prices are likely to become part of the economic landscape.[5]

The widening gap between the demand and the supply of grain can be seen in nearly all the more populous developing countries. Two years ago, in *Full House,* we projected the grain supply/demand balance to 2030 for 11 of the largest developing countries, home to some two thirds of humanity. (See Table 7-2.) If the anticipated growth in population for these countries materializes by then, the need for imports everywhere except Brazil would increase far beyond the current level even if there is no improvement in diets.[6]

China, for example, which was essentially self-sufficient in 1990, would need to import some 215 million tons of grain in 2030. India would need 45 million tons of grain from abroad, not nearly as much as China because India is a semivegetarian society, requiring less grain for feed. These two and nine other developing countries, some of whose populations are projected to more than triple, would see their combined need for imported grain climb from 38 million tons in 1990 to

TABLE 7-2. *Grain Imports for Selected Countries,
1990, With Projected Need for Imported Grain in 2030*

Country	1990	2030
	(million tons)	
Bangladesh	1	9
Brazil	6	4
China	6	215
Egypt	8	21
Ethiopia and Eritrea	1	9
India	0	45
Indonesia	3	12
Iran	6	32
Mexico	6	19
Nigeria	0	15
Pakistan	1	26
Total	38	407

SOURCE: See endnote 6.

more than 400 million tons in 2030.[7]

All these countries face a scarcity of cropland. Many, including the most populous countries—China, India, and Indonesia—will lose this land as they industrialize. In some, future gains in food production will be severely constrained by water scarcity. Among them are Iran, Pakistan, and Mexico. For others, the sheer magnitude of projected population growth will simply overwhelm their agricultural systems. Nigeria, whose population is projected to triple, is in this category.[8]

Brazil is one of the countries with the largest potential for expanding production. But even with a hefty growth in its grain output, it will still find itself scrambling to remain self-sufficient. Indeed, in recent years Brazil has been losing the battle to be self-sufficient in grain, becoming the largest importer in the western hemisphere. Beyond this, it is projected to add nearly 100 million people by 2030, reaching a total of 252 million, roughly the same size as the United States today. Feeding another 100 million people while its population simultaneously moves up the food chain, consuming more livestock products, will probably take all the agricultural resources Brazil can mobilize.[9]

Among regions, the growing deficit in Asia stands out. In 1950, Asia's net grain imports totalled 6 million tons, far less than the 22 million tons imported in Europe, the leading grain-deficit region at the time. But by 1995, Asia's net imports had climbed over 90 million tons, making it far and away the dominant importing region. By 2030, its import needs could easily exceed 300 million tons. This is not surprising since the region contains more than half the world's people, is responsible for 60 percent of world population growth, and has the fastest economic growth of any region in history.[10]

On the other side of the global import/export equa-
tion, North American grain exports between 1950 and
1995 climbed from 23 million to 116 million tons, with
the United States accounting for 94 million tons and
Canada 22 million tons. There is reason to doubt
whether over the next 35 years the United States (projected to add some 82 million people), Canada, smaller
exporters—such as Australia, Argentina, South Africa,
and Thailand—and some potential new exporters in
Eastern Europe, all small ones, can come close to match-
ing the projected need for imports in Asia, Africa, and
the other importing regions at traditional prices. In contrast to projections by the U.N. Food and Agriculture
Organization and the World Bank of declining grain
prices, this analysis suggests that substantial price rises
will be needed to balance demand and supply.[11]

The prospect in the 11 countries in Table 7-2 con-
trasts sharply with that in Europe, where the growth in
both population and the consumption of livestock products has come to a halt. With grain use stable or declining since 1980 and with production outpacing it since
the early eighties, Europe has developed an exportable
surplus of grain rivalling that of Canada, the world's second ranking grain exporter after the United States. (See
Figure 7-1.) It is the only geographic region where the
demand for grain has levelled off, and it has done so
within the carrying capacity of its land and water
resources.[12]

Of course, the demand for grain and its supply always
balance in the marketplace. But at what price? And
what will be the economic and social effects of moving to
that higher price level? Between early 1994 and mid-
1996, the export price of wheat from the United States
climbed from just over $4 a bushel to more than $7 a

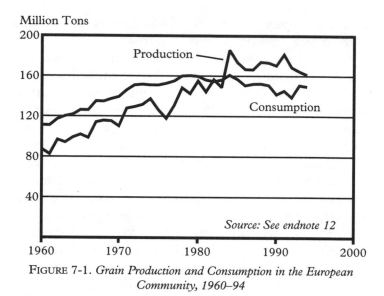

FIGURE 7-1. *Grain Production and Consumption in the European Community, 1960–94*

bushel. No one knows for sure what the price is likely to be over the longer term, but all indications are that it is going to be much higher than that which prevailed before this recent climb.[13]

The world appears to be moving into a new era, one in which the problems governments face will be vastly different from those that now preoccupy them. Given the prospect of rising food prices and the wide disparities in income among countries, price rises that are merely inconvenient for some may become life-threatening for others.

If the rising food prices that are in prospect materialize, the terms of trade among geographic regions will be altered dramatically. Among the continents, the winners will be North America, Europe, and Australia, the regions with grain surpluses. The big winner in immediate economic terms, of course, will be North America, since the United States accounts for nearly half of world grain exports and Canada, for another 10 percent. With population now essentially stable throughout Europe and with limited prospect for further movement up the food chain, future growth in grain output in this region will convert largely into exportable surpluses. In addition to Western Europe, countries such as Poland, Romania, and the Ukraine, where populations are also stable, may have exportable surpluses of grain.[14]

Asia, with its unprecedented rate of industrialization and dense population, will be the loser in the sense not only that its dependence on imported grain is rising at such a spectacular rate, but also because the price of this grain will be rising. Higher prices will mean steadily growing outlays for imported grain for both food and feed. Asia is becoming industrially strong, but agriculturally vulnerable.

Africa, of course, will be the big loser. Its demand for grain, driven by the fastest population growth of any continent in history, is projected to continue to outstrip the growth in production, leaving the region heavily dependent on imports that it may not be able to afford.

In rural/urban terms, the countryside will be the winner and the cities, the losers. Indeed, emerging food scarcity may reverse the terms of trade between the countryside and the city. Land and water values seem certain to escalate. Since the beginning of the Industrial Revolution, cities—with their control of capital and technology—have had an advantage. But in a world where capital, technology, and certainly labor are relatively abundant and where land and water are scarce, a reversal in the terms of trade may be inevitable.

The new era brings with it a new political dynamic. In the past, as long as the pie was expanding rapidly, political leaders could urge patience, arguing that everyone's lot would soon improve. But when the pie stops expanding—not because of a temporary lag of technology or planning, but because our collective consumption has finally run into some of the earth's natural limits—the political dynamic changes. How the pie is sliced takes on a new prominence.

The lagging production and the falling grain stocks of recent years suggest that the politics of surplus, which dominated the last half-century, will be replaced by a politics of scarcity. Instead of a few exporting countries competing for markets that were never quite large enough, some 120 importing countries will compete for supplies that never seem adequate. Grain export subsidies may be replaced by export taxes or even outright embargoes to keep food prices at home down. Conflict among countries over access to fisheries will become commonplace. Struggles over the control of water will intensify among national governments, between those in the upper and lower reaches of river watersheds, and between cities and farmers.

Governments of grain-exporting countries, sensitive to the speed with which the vagaries of a global market can generate food scarcities and inflation at home, will be tempted to impose export embargoes. Indeed, in May 1995 Vietnam imposed a partial embargo on rice. Because grain prices in neighboring China had risen well above the world market level, large amounts of rice from Vietnam were crossing the border. But with rice prices climbing by up to 70 percent in northern Vietnam, making it difficult to control inflation, the government restricted exports while waiting for the new harvest. Since Vietnam is the third largest rice exporter (after Thailand and the United States), this raised the world rice price. Meanwhile, some lower-income inland provinces in China have even banned grain shipments to the more prosperous coastal provinces in an effort to stem price rises.[15]

As world grain prices escalated in the late summer and fall of 1995, leaders within Europe became concerned that this would translate into rising food prices. In early December 1995, the European Union imposed a wheat export tax of $32 a ton in an effort to arrest the rise in bread prices. In early January 1996, a similar step was taken for barley, the principal feedgrain, to control prices of livestock products. In effect, with these actions a two-tiered grain price system was created in the world: a lower price within Europe and a higher price outside it, where the world's low-income populations live. With the stroke of a pen, the European Union's agricultural policy effectively nullified the positive effects of its international food assistance programs.[16]

The prospect of chronic world food scarcity raises new questions about the morality of restricting or banning food exports. Just as the international community once wrestled with the question of whether there are any conditions under which exporting countries are justified in subsidizing exports, it must now decide whether there are any circumstances under which a government is justified in restricting exports in order to quell domestic food price rises, even though it will lead to more rapid price rises in the rest of the world. The agricultural part of the General Agreement on Tariffs and Trade negotiations, from the Kennedy through the Uruguay Rounds, focused on ensuring exporting countries' access to markets. The challenge now is to devise a set of trade institutions and rules that will ensure importing countries' access to supplies.

In the world of the late nineties, the number of grain-importing countries will dwarf that of exporters. Only a handful of countries now consistently export grain on a meaningful scale: Argentina, Australia, Canada, France, Thailand, and the United States. Current world grain exports add up to roughly 200 million tons per year, of which the United States accounts for close to half, controlling a larger share of world grain exports than the Saudis do of oil. This puts great power in the hands of one government, raising the possibility that food could be used for political purposes.[17]

The politics of scarcity is not confined to grain alone. Now that the demand for seafood exceeds the sustainable yield of fisheries, those managing this resource must determine what the sustainable yield of a fishery is, negotiate the distribution of that catch among competing interests, and enforce adherence to any quotas established. Where fisheries are shared among countries, as is often the case, the process becomes infinitely more complex.

With seafood, the politics of scarcity can be seen in the increasingly frequent clashes among countries over access to fisheries. These include cod wars between Norwegian and Icelandic ships, between Canada and Spain over turbot off Canada's eastern coast, between China and the Marshall Islands in Micronesia, between Argentina and Taiwan over Falkland Island fisheries, and between Indonesia and the Philippines in the Celebes. The pervasiveness of these conflicts is evident in a Greenpeace statement released at the U.N.-sponsored conference on regulating fishing on the high seas in July 1995: "Tuna wars in the northeast Atlantic, crab wars in the North Pacific, squid wars in the southwest Atlantic, salmon wars in the North Pacific, and pollock wars in the Sea of Okhotsk are all warning signs that fish stocks are in serious trouble."[18]

Conflicts over water are also becoming commonplace. India and Bangladesh cannot agree on the allocation of Ganges waters. The United States and Mexico compete for the waters of the Rio Grande. Egypt and the Sudan, which had earlier agreed to distribute the waters of the Nile between them, are now engaged in new negotiations that include Ethiopia and other countries in the upper Nile watershed. Israel, Jordan, and representatives of Palestine are trying to implement recently signed agreements allocating water from the Jordan River and other shared water resources. To the north, Turkey is preempting the waters of both the Tigris and the Euphrates rivers at the expense of downstream countries, Syria and Iraq. The Central Asian countries that use water from the Amu Dar'ya and the Syr Dar'ya must decide whether they want to save the Aral Sea and, if so, how to allocate the water of the rivers in such a way that enough is left to sustain the sea. As the value of land and water climbs in the years ahead, as now seems likely, the competition over these finite resources is certain to intensify.[19]

Increasingly, the constraints the world is facing are environmental, not economic. Historically, the size of the fish catch was largely determined by the investment in fishing trawlers, but today the sustainable yield of fisheries is the controlling factor. Until recently, the amount of water pumped was determined by the number of wells drilled, but now it is the sustainable yield of the aquifer. And, increasingly, it is not the amount of fertilizer that a farmer can afford but the amount of nutrients that plants can absorb that dictates grain production levels.

At a time of growing scarcity and high grain prices, the need for international food assistance is rising, but many aid donor countries, including the United States, have slashed food assistance budgets during the nineties. These cuts, combined with higher procurement prices for grain, have reduced food aid from its historical high of 15.2 million tons of grain in fiscal 1993 to an estimated 7.6 million tons in 1996. Just when the need for food assistance is rising, food aid is diminishing. When the problem facing exporting countries was surpluses, it was easy to justify generous budgets for food assistance, because they simultaneously reduced surpluses and alleviated hunger. But when the allocations of grain for food assistance begin to compete with domestic consumers and to drive prices up, as seems likely, it will be far more difficult to be unselfish.[20]

Food security in the decades ahead goes far beyond access to food aid for low-income, food-deficit countries to include access to supplies by all food-importing countries. For the 120 or so countries that import grain, the years ahead are fraught with risk. Apart from the overall scarcity and higher prices that will result if import needs continue to outpace export supplies, higher prices also induce export restrictions by the exporting countries, which only exacerbates the scarcity. With the recent imposition of restrictions on rice exports by Vietnam, the world's third ranking rice exporter, and restrictions on both wheat and feedgrain exports by the European Union, which rivals Canada as a grain exporter, there is no reason to assume that other export restrictions will not follow.

In a world of scarcity, the risk of export restrictions, including outright embargoes is all too real. What if the United States were to follow Europe and restrict exports of grain, or even ban them, as it did with soybeans in 1973? How much would food prices have to rise in the

United States before consumers would begin to press for export limits or an export embargo? At what point on the food price index would rising prices lead American consumers to resist sharing U.S. grain with China's increasingly affluent 1.2 billion consumers?[21]

Beyond this, the world's importing countries are at risk because of heavy dependence on a single geographic region and climate change. For 120 countries to be dependent for the bulk of their grain imports on two countries that are affected by the same climate cycle further heightens the risk of import dependence. If the United States experiences severe heat and drought in the wheat-growing Great Plains, Canada often does as well.

With the expansion of irrigation in both China and India over the last few decades, harvest levels now fluctuate relatively little from year to year. But in the United States, which has a largely rain-fed grain culture, year-to-year swings can easily increase or decrease the grain harvest by one fifth, making it the principal source of variations in world output. Over the last decade, the U.S. corn crop, which normally accounts for one eighth of the world grain harvest, has ranged from a low of 125 million tons in 1988 to a high of 257 million tons in 1994. A year-to-year drop of one third in the U.S. corn harvest is not unusual. Annual changes in the U.S. grain harvest can thus increase or decrease world output by as much as 5 percent.[22]

For the world as a whole, the 1995 grain harvest dropped some 60 million tons below the initial estimate largely because of crop-withering heat waves in the temperate countries of the northern hemisphere during the summer growing season. With atmospheric carbon dioxide levels now rising steadily, forecasting longer-term climate patterns is becoming increasingly difficult.[23]

The only safe policy for grain-importing countries is to avoid becoming heavily dependent on imports. The volatility of national politics in exporting countries and the growing uncertainty about future climate patterns combine to argue for a strenuous effort to try to stabilize population within a country's food carrying capacity. For low-income grain-importing countries, the threat of higher prices is not something to be taken lightly.

Grain export restrictions over the last two years in Vietnam and the European Union remind us that in times of scarcity, access to supplies cannot be taken for granted. The record grain prices set in the spring of 1996 serve as warning that affluent industrial societies will not hesitate to bid grain prices up to at least twice their historical level to satisfy their needs for food and feed, regardless of how it affects the rest of the world.[24]

The politics of surplus that dominated the last half-century in agricultural trade relations was troublesome, and a thorn in the relationship among exporting countries as they tried to outdo each other with export subsidies, both direct and indirect. But the politics of scarcity is more than troublesome. It is dangerous. Export embargoes or restrictions at a time of acute scarcity can threaten lives and topple governments. In addition to the sensitive issue of international access to grain supplies, access to fisheries and to water resources will further escalate the politics of scarcity, putting new pressures on governments and international institutions.

8

Responding to the Challenge

If food scarcity becomes chronic and food price rises become uncontrollable at times of acute scarcity, it will be a sure sign that our global economic system is becoming environmentally unsustainable. In such a scenario, it is not merely the security of the world's food supply that is at stake, but the stability of civilization.

Understanding what is happening depends on our ability to analyze the fast-changing relationship between the earth's environmental system and our economic and political systems. Analyzing each of these in its own right is a challenge, but to analyze and project the ongoing interaction among the three verges on the impossible. We can say, however, with more confidence than we would like, that if environmental degradation proceeds far enough, it will translate into economic instability in the form of rising food prices, which in turn will lead to political instability.

If the economy is no longer able to respond readily to price signals because it is colliding with some of the earth's natural limits, then governments may be forced to formulate a new strategy to achieve a humane balance between food and people. Stated otherwise, if fishers and farmers cannot rapidly expand the supply, governments may have to focus their efforts on slowing the growth in demand. At a minimum, this may mean stabilizing population size in many countries much sooner than political leaders have anticipated and much sooner than has been projected.

As noted in Chapter 1, emerging food scarcity is the first major manifestation of an environmentally unsustainable global economy. The steps now needed to ensure future food security are precisely the ones needed to build an environmentally sustainable economic future. In effect, food scarcity provides a powerful rationale for environmental action.

Prominent among the formidable tasks ahead are stabilizing population and climate. Other challenges include protecting soils on the world's croplands by reducing erosion to the rate of new soil formation, stabilizing aquifers by reducing pumping to the rate of recharge, and securing the output of fisheries by reducing the catch to the sustainable yield. If the future matters, we have no choice but to restore the natural balances and stability that prevailed throughout most of our existence as a species.

It is hard to find any historical precedent for a challenge on the scale that the world is now facing. The only remotely comparable situation would be the mobilization of resources and people and the restructuring of economies that was associated with World War II. But there are differences. Once the war was over, things returned to normal. In contrast, what is needed now is a permanent restructuring of the global energy economy, shifting from a fossil fuel-based one to a solar/hydrogen economy. And stabilizing population depends on a lasting shift in our reproductive behavior—one that will lead to smaller families everywhere.

The challenge is particularly difficult because some of the actions needed are economically counterintuitive. While the ecology of sustainability argues for reducing the amount taken from most oceanic fisheries, the economic indicators argue for increasing the catch. Traditionally, higher prices were a signal to invest in more trawlers; now this simply leads to the collapse of more fisheries. And the higher prices rise, the more difficult it becomes for governments to limit catches to the ecologically sustainable level.

A similar situation exists with water. When water tables are falling as a result of excessive demand, aquifer stabilization requires a cutback in pumping. But with the demand for water also growing each year, and with water scarcity spreading, the price of water is rising, arguing for an increase in pumping—not a reduction. These are just a few reasons that building a sustainable future is so difficult.

Regaining control of our destiny will require some tough choices—some of the most difficult ones that societies have ever had to make. Even given the full exploitation of agricultural technology, the land and water constraints facing many developing countries will make it impossible for them to provide for both the population growth now projected and the diversification of diets so widely desired. More broadly, our generation will have to choose between reducing fossil fuel use for ourselves and the prospect of reduced food security for our children. In water-scarce regions, governments will have to choose between water for indoor plumbing and water for irrigation. Densely populated countries, such as China, will have to decide whether to use land for automobile-centered transportation systems or for meeting food needs. This is but a small sampling of the tough choices that lie ahead.

Historically, balancing the supply and demand for food was largely the responsibility of agricultural ministries and farmers, a matter of adjusting farm policies and investing more in agriculture. But now the challenge is far more complex, engaging all of society. It involves childbearing decisions, energy policies, and dietary habits as well as agricultural policies. If ministers of agriculture are smart, they will be meeting with their family planning counterparts in government, explaining to them that achieving a humane balance between food and people in their country may now depend more on slowing population growth than on expanding the fish catch or the grain harvest.

Stabilizing world population will require nothing less than a revolution in human reproductive behavior. The rise in living standards that moved some 31 industrial societies containing some 14 percent of the world's population through the demographic transition, leading to a balance between births and deaths, is not proceeding fast enough to stabilize population size in most countries before the natural limits of life-support systems are reached. The challenge is to achieve stability quickly in the countries with the remaining 86 percent of humanity, because otherwise there may be little hope of upgrading diets.[1]

In a world where both the seafood catch and grain output per person are falling, it may be appropriate to ask whether there is any moral justification for having more than two children per couple, the number needed to replace ourselves. It may be time for the world's leaders, including the Secretary-General of the United Nations, the President of the World Bank, and national political leaders, to speak out, urging couples everywhere to draw the line at two surviving children per couple.

The good news is that contrary to conventional wisdom, it is possible to slow population growth quickly. Perhaps the most impressive example is the decline in Japan's population growth after World War II. In the late forties, following the loss of its occupied territories in Southeast Asia and the associated loss of access to the oil, minerals, rubber, forest products, and rice of the region, Japan was forced to devise a strategy to survive on the resource-poor chain of mountainous islands that make up its land base.

The abrupt change in its circumstances that came with its defeat in 1945—a change not unlike the one the world is facing today on the food front—led to some basic rethinking. Among other things, the government abandoned the pronatalist policy of the thirties and early forties, replacing it with a policy designed to slow population growth. Between 1949 and 1956, this policy shift, combined with economic recovery, cut the population growth rate in half. Lacking modern contraceptives, Japan built its family planning program around the use of condoms backed up by abortion.[2]

A similar situation developed in China at the beginning of the seventies. After insisting for more than two decades that rapid population growth was an asset rather than a liability, the leaders in Beijing concluded that if

China stayed on its current population trajectory it would face an eventual decline in living standards as population growth overwhelmed local life-support systems. They adopted a policy of later marriage and a social goal of no more than two children per couple. Later in the seventies, realizing they had waited too long to slow the growth of population, the Chinese were forced to adopt a goal of one child per couple. As a result of the new policy adopted in the early seventies, China was able to cut its population growth in half between 1970 and 1977.[3]

China has been widely criticized for its one-child-family policy. But given the limits on the population carrying capacity of its land, water, and forest resources and the aspirations of its people for a better life, it had no choice. To its credit, it has moved more people out of poverty in a shorter period of time than ever before in history.

More recently, the government of Iran has begun to worry about population growth. After the downfall of the Shah in 1979, the Ayatollah Khomeini argued that Iran needed "babies, lots of babies" to enable the Islamic republic to triumph over the "degenerate West." But when the 1986 census was taken, it revealed a population of 50 million, much larger than the 36 million referred to by the Ayatollah in his speeches. Recognizing the difficulties that lay ahead on this demographic path, which would lead to a population of 183 million by 2030, the Islamic government initiated an ambitious family planning program, bolstered by the support of religious leaders who publicly went on record favoring contraception and smaller families.[4]

This family planning program, launched in 1989, encouraged a minimum of three to four years between pregnancies, discouraged pregnancies of women younger than 18 and older than 35, and urged a limit of three children per family. The organization of this national program, including the extension of services throughout the countryside and the provision of free oral contraceptives, condoms, and female and male sterilization, has contributed to a decline in the number of births per woman from 7.0 in 1986 to an estimated 5.1 in 1996.[5]

In 1993, the government of Iran reinforced its urgings for smaller families by announcing that as of 1994, the subsidies for health care, housing, and education that traditionally came with each birth would be limited to three children per couple. The elimination of any social subsidies beyond the third child has an economic effect on childbearing decisions, but perhaps even more important, it indicates that the government considers rapid population growth to be a serious threat to the society. This raises the question of whether other governments should continue to offer tax deductions to couples for unlimited numbers of children, as, for example, the United States still does.[6]

The Iranian program is in many ways a model since it is strongly supported from the top by national leaders; it is organized at the grassroots level, with a local volunteer being responsible for roughly 50 families; it offers a full range of contraceptive services; and it succeeded in increasing the share of girls 15 to 19 years of age enrolled in school from 25 percent in 1986 to 54 percent in 1993.[7]

The dramatic policy turnabout in Iran came none too early, given the deterioration in life-support systems. Water tables are falling. The forests have largely disappeared. Overgrazing of rangelands is widespread. Soil erosion threatens cropland productivity. Under these conditions, there was little hope for future generations if Iran's population tripled. In effect, concerns about ecology overrode those of theology, leading to a basic shift in population policy.

Beyond the abandonment of pronatalist policies, the most immediate need on this front is to fill the family planning gap—getting services to the estimated 120 million women in developing countries who want to limit their families but lack access to the means to do so. In addition, there is a need to provide equal educational and economic opportunities for women. This is the key to the elimination of the gender bias that pervades so many societies. As the educational level of women rises, family size declines. This correlation cuts across every culture. Future food security and political stability may depend more on the education of young females in developing countries than on any single investment other than in family planning itself.[8]

Perhaps the most neglected population policy issue is the lack of information on the relationship between population size and dietary aspirations on the one hand and the carrying capacity of national food support systems on the other. An assessment of a country's food systems—fisheries, rangelands, and croplands—can help people understand the difficult choices they face between moving up the food chain and having more children. The desire to have many children is strong, but so is the desire to diversify diets. Tough choices will have to be made.

Closely rivaling population stabilization in degree of difficulty is stabilizing climate. Both the technologies and the economics are now falling into place for the shift from a fossil-fuel based energy economy to one based on renewable energy sources. In 1995, for example, coal and oil production expanded by roughly 1 percent each. But wind electric generation expanded by 33 percent. And the sales of photovoltaic cells grew by 17 percent.[9]

The potential for wind energy is enormous. Wind electric generation in California is now sufficient to meet the residential needs of the city of San Francisco. Three U.S. states—North Dakota, South Dakota, and Texas— together have enough harnessable wind energy to meet national electricity needs. In Europe, wind power could theoretically satisfy all the continent's electricity needs. And sizable wind generation projects are under way in China to help that country meet its growing demand for power. Today the world gets one fifth of its electricity from hydropower, but this is dwarfed by the potential of wind power.[10]

The use of photovoltaic cells, which convert sunlight into electricity, is now expanding rapidly in the Third World. In remote or widely dispersed villages where there is no electric grid in place, it often costs less to install solar cells than to build a centralized power plant and a grid to deliver the electricity from it. Both national governments and international lending agencies, such as the World Bank, are now encouraging the development of this new source of energy.[11]

Once cheap electricity becomes available from wind power, solar thermal power plants, or other low-cost renewable sources, that electricity can be used to electrolyze water, producing hydrogen, which provides a convenient way of both storing and transporting solar energy (including wind) in its various forms. One of the exciting things about the rapid growth in natural gas, which is both less polluting and emits less carbon per unit of energy produced than other fossil fuels, is that it can serve as a transition fuel from the fossil-based ener-

gy system to a renewable-based one. The infrastructure
for natural gas can easily be adapted for hydrogen. The
combination of electricity and hydrogen can meet all the
energy needs of a modern industrial economy, from
transportation to computers.[12]

One of the keys to accelerating the transition to
renewable energy is a carbon tax that would reflect the
cost to society of burning fossil fuels—air pollution, acid
rain, increasingly destructive storms, and ever more
intense crop-withering heat waves. Data collected by the
insurance industry indicate a steep rise in property damage in recent years as a result of more destructive storms.
They show that claims from weather-related disasters
worldwide have increased from $16 billion during the
eighties to $48 billion during the first half of the nineties.
These increases have left the industry in a state of shock,
converting some of its leaders into proponents of policies
that reduce carbon emissions.[13]

No one has tried to make a similar calculation for the
cost to farmers of harvest losses as a result of the increasingly intense heat waves associated with the rise in average temperature since 1979. We do know that the entire
northern tier of industrial countries suffered from crop-
damaging heat during the summer of 1995. These heat
waves, such as the one that claimed 465 lives in Chicago
in July 1995, helped drop the world grain harvest to its
lowest level since 1988.[14]

As these costs of rising global temperatures are more
clearly identified, the pressure by the insurance industry,
and quite possibly by farmers, to reduce dependence on
fossil fuels will intensify.

In addition to stabilizing population and climate, policymakers need to protect the topsoil on the world's
cropland, since every ton of topsoil lost to erosion today

diminishes the food supply for the next generation. Ensuring food security now depends on systematically devising a world plan of action to stabilize soils on erodible cropland, much as the United States has done with its Conservation Reserve Program. (See Chapter 4.)

Future food security also depends on protecting cropland from conversion to nonfarm uses. Perhaps the most effective way to do this is to adopt a stiff cropland conversion tax, one that would be large enough to reflect the land's long-term contribution to food security. This would at least force those planning to use highly productive cropland for industrial or residential construction to weigh alternative sites seriously.

As the dimensions of cropland scarcity begin to manifest themselves in rising food prices, the land intensity of transportation systems is getting more attention. As noted in Chapter 4, this debate is surfacing in China, which decided in 1994 to develop an automobile-centered transportation system, one patterned after those in industrial countries. Since the oil price hikes of the seventies, it has been fashionable to worry about the energy needed to supply an ever-growing world automobile fleet. But the more serious constraint may be the land required for a transport system based on private automobiles. For countries like China and India, concerns about future food security argue for replacing their auto-centered transportation strategies with one built around a state-of-the-art rail passenger system combined with bicycles. This would maximize mobility for the entire society while minimizing cropland losses.[15]

While governments and agricultural researchers have focused on increasing land productivity over the past half-century, efforts to increase water productivity have been largely neglected. In part, this was because there were no markets in water as there were in land, so there was no strong economic motivation to research technologies that would increase efficiency. Now that water scarcity is spreading, the return on investment in new water technologies, in more-efficient irrigation practices, and in the shift to more water-efficient crops is climbing.

Even as better irrigation technologies are being devised, farmers can begin to shift to more water-efficient crops. It takes less water to produce a ton of wheat than a ton of rice. Where environmental conditions permit this shift to be made without losing yield, the emergence of water markets would encourage the planting of water-efficient crops.[16]

Partly as a result of the excess production capacity and depressed commodity prices of the eighties and early nineties, and partly as a result of World Bank and U.N. Food and Agriculture Organization projections of continued surplus capacity, investment in agricultural research lagged. U.S. funding for the international network of some 15 leading organizations in developing countries fell sharply. This funding should be restored and indeed expanded well beyond the current level, since this network provides a means not only of developing new technologies but also of disseminating these technologies internationally.[17]

Investing more in agricultural research will not necessarily lead to a new technology that will produce a quantum jump in world food production. But every technological advance, however small its contribution, buys additional time in which to stabilize world population size. In the absence of the enormous contribution to growth in the world grain harvest that came from expanding fertilizer use, the development of new technologies to expand production is more important than ever. This demands an all-out effort using both conventional plant breeding techniques and biotechnology.

While pushing hard to stabilize population size, the world can return the small amount of remaining cropland idled under commodity supply management programs to production. As noted in Chapter 4, in late 1995 the U.S. Department of Agriculture announced that it would release for production the small amount of remaining set-aside land under its commodity programs. The European Union (EU) declared it would reduce its set-aside from 12 percent of its cropland base in 1995 to 10 percent in 1996. If the EU returns the remaining 10 percent of this idled grainland, it could add another 18 million tons to the world grain harvest, enough to cover world population growth for eight months.[18]

After commodity set-aside, the only remaining land reserve is that in the Conservation Reserve Program in the United States, which converted some 14 million hectares of cropland, much of it highly erodible, to grassland. At least half of this could be returned to use and farmed sustainably if the appropriate cropping practices were adopted, such as minimal tillage. Bringing this land, which is predominantly wheatland in the Great Plains, back into production could boost the world grain harvest by perhaps another 24 million tons, enough to cover 10 months of world population growth.[19]

Another way to expand the food supply is to convert land now used for nonessential, nonfood products to food production. For example, some 10 million tons of corn was used in the United States in 1995 to produce fuel-grade ethanol. Phasing out this use of corn would release enough grain for food consumption to support world population growth an additional four months. This shift, in fact, began in 1996 as the price of corn

climbed to record highs. Some distilleries were closed in
1995; others are certain to be closed if corn prices continue to rise over the longer term.[20]

If the international community ever got serious about
phasing out tobacco to protect health, some 5 million
hectares of cropland now used to produce this crop
could be converted to the production of grain. This
would provide enough to support world population
growth for six months, while simultaneously reducing
health care expenditures and freeing up capital for
investment in agriculture.[21]

Persuading half the people in the world to shift from
cotton to synthetic fibers, at least as long as the oil is
available to produce them, would release 14 million
hectares of land currently in cotton. Putting these fields
into grain would yield 50 million tons of grain, covering
17 months of world population growth.[22]

In a world where grain stocks are dangerously low—a
situation that may continue for some time—and where
idled cropland is being returned to production, the only
remaining major reserve is the 36 percent of the world
grain harvest—some 640 million tons—that is fed to
livestock, poultry, and fish. There are several ways to tap
this reserve to buy time to stabilize population.[23]

One way is to let the market do it. As grain prices rise,
the world's affluent will move down the food chain.
When grain prices doubled in the seventies, Americans
lowered their consumption of meat, milk, and eggs
enough to reduce grain feeding from 143 million tons in
1973 to 101 million tons in 1974—a drop of 42 million
tons. The disadvantage of this approach is that when
prices rise high enough to move the affluent down the
food chain, they may be life-threatening to the world's
poor.[24]

Another approach would be for governments to educate people about the health risks associated with excessive consumption of fat-rich livestock products. The healthiest people in the world are not those living at the top of the food chain or those at the bottom, but those in the middle. Italians who use about 400 kilograms of grain per year, for example, live longer than Americans who consume twice that.[25]

During World War II, a number of industrial countries rationed the consumption of livestock products. The disadvantage of this approach is that it requires a nationwide bureaucracy to administer the program and to enforce compliance.

But perhaps the most efficient technique for moving affluent consumers down the food chain would be a tax on the consumption of livestock products, one not unlike that applied by many governments to grain-based alcoholic beverages. Although this might not be politically popular among the affluent, it would moderate grain price rises. And at a time of acute scarcity, it could be the price of political stability.

Reducing the consumption of livestock products, through whatever means, could help buy some additional time to stabilize population size. If the world's affluent cut their consumption of grain-fed livestock products by 10 percent, they could free up 64 million tons of grain for direct human consumption, enough to cover world population growth for some 27 months.[26]

Both market forces and government policies in many countries are encouraging a shift from the less efficient forms of converting grain into livestock, such as cattle in feedlots, to the more efficient ones, such as broiler production. Market forces are already shifting the production pattern of livestock products, rapidly increasing

broiler production while discouraging additional investment in cattle feedlots. By the end of this decade, poultry meat production is likely to overtake that of beef,
including both the beef produced on rangelands, the
dominant share, and that produced in feedlots.

★   ★   ★   ★

Clearly, securing adequate food supplies in the years
ahead is not just a matter of fine-tuning farm policies or
investing more in agriculture, important though these
steps are. Instead, it involves reversing the trends that
are destroying the economy's environmental support systems.

Whether we succeed in this vast undertaking depends
on how much we care about the next generation. Are
we, as individuals, willing to work for the changes needed to get the world economy off the path leading toward
food scarcity, economic instability, and social disintegration? Up until now, most people have treated efforts to
ensure an environmentally sustainable future like a spec-
tator sport: they have been sitting in the stands watching
a handful of active participants on the playing field. But
now we have an economic rationale for turning around
the demographic and environmental trends of recent
decades, because if we do not act quickly, the next generation may go hungry.

Political action is needed to reshape national policies
that affect such things as population growth and the
earth's climate. In the spring of 1996, for example, just
when wheat and corn prices were reaching all-time
highs, the U.S. Congress was cutting international family planning assistance programs by some 85 percent.
Some individuals concerned about the effect this would

have on world population growth contacted members of Congress to protest this move. Public interest population groups worked to restore this funding, but they were not able to mobilize enough support to do so. Unless those of us who care about the future are willing to work for it, the world our children inherit may not be the one that we want for them.[27]

The keys to a sustainable future are information and action. First we educate ourselves on the issues, then we act. Most countries today have democratically elected governments—governments that can stay in power only if they respond to the concerns of their people. In democracies, responsibility for governance lies with the individual. Change comes about because people care enough to work for that change. Otherwise, the status quo and the special interests prevail.

If food scarcity is emerging as the defining issue of the new era now unfolding, it will profoundly challenge political leaders everywhere. History judges leaders by whether they respond to the great issues of their time, such as slavery in the United States during the last century or the rise of fascism in Europe in this century. For us, food security may be the great issue of our time.

Notes

CHAPTER 1. The Challenge

1. "Grain Prices Continue to Climb; Official Urges Calmer Trading," *New York Times*, April 26, 1996; "Prices for Wheat and Corn Drop on Hopes for Improved Harvests," *New York Times*, April 30, 1996; "Futures Prices," *Wall Street Journal*, various editions.
2. U.S. Department of Agriculture (USDA), "Production, Supply, and Distribution" (electronic database), Washington, D.C., February 1996.
3. Donald O. Mitchell and Merlinda D. Ingco, International Economics Department, *The World Food Outlook* (Washington, D.C.: World Bank, 1993); U.N. Food and Agriculture Organization (FAO), *World Agriculture: Towards 2010* (New York: John Wiley & Sons, 1995).
4. Annual population increase from U.S. Bureau of the Census, Center for International Research, Suitland, Md., private communication, January 23, 1996.
5. Anjali Acharya, "CFC Production Drop Continues," in Lester R.

Brown, Christopher Flavin, and Hal Kane, *Vital Signs 1996* (New York: W.W. Norton & Company, 1996).

6. USDA, op. cit. note 2; FAO, *Yearbook of Fishery Statistics: Catches and Landings* (Rome: various years).

7. Bureau of the Census, op. cit. note 4; USDA, Foreign Agricultural Service (FAS), *Grain: World Markets and Trade,* Washington, D.C., June 1996.

8. USDA, op. cit. note 2.

9. Joel E. Cohen, "How Many People Can Earth Hold?" *Discover,* November 1992.

10. FAO, op. cit. note 6; Bureau of the Census, op. cit. note 4.

11. Information on oceanic fish catch from FAO, op. cit. note 6; information on fisheries decline based on statistical data in the FAO fisheries database, FISHSTAT-PC, FAO Fisheries Statistics Division, Rome, 1994.

12. USDA, "World Grain Database" (unpublished printouts), Washington, D.C., 1991; USDA, op. cit. note 2; USDA, "World Agricultural Supply and Demand Estimates," Washington, D.C., January 1996; USDA, FAS, *Grain: World Markets and Trade,* Washington, D.C., January 1996.

13. USDA, op. cit. note 2; USDA (unpublished printout), op. cit. note 12; USDA, "World Agricultural Supply and Demand Estimates," Washington, D.C., January 1995.

14. Sandra Postel, "Forging a Sustainable Water Strategy," in Lester R. Brown et al., *State of the World 1996* (New York: W.W. Norton & Company, 1996).

15. World grain production from USDA, op. cit. note 2; irrigated area from FAO, *FAO Production Yearbook 1994* (Rome: 1995).

16. Gary Gardner, "Preserving Agricultural Resources," in Brown et al., op. cit. note 14.

17. USDA, Natural Resources Conservation Service, *Summary Report: 1992 National Resources Inventory* (Washington, D.C.: July 1994, rev. January 1995); Sandra Postel, *Last Oasis: Facing Water Scarcity* (New York: W.W. Norton & Company, 1992); USDA, op. cit. note 2.

18. FAO, *Yield Response to Water* (Rome: 1979).

19. FAO, *Fertilizer Yearbook* (Rome: various years); K.F. Isherwood and K.G. Soh, "Short Term Prospects for World Agriculture and Fertilizer Use," presented at 21st Enlarged Council Meeting, International Fertilizer Industry Association, Paris, November 15-17, 1995.

20. USDA, op. cit. note 2.

21. James Hansen et al., Goddard Institute for Space Studies Surface Air Temperature Analyses, "Table of Global-Mean

Monthly, Annual and Seasonal Land-Ocean Temperature Index, 1950-Present," as posted at http://www.giss.nasa,gov/Data/GIS-TEMP, January 19, 1996.

22. USDA, op. cit. note 2.
23. International Monetary Fund (IMF), *World Economic Outlook, October 1995* (Washington, D.C.: 1995); population information from Population Reference Bureau (PRB), *1995 World Population Data Sheet* (Washington, D.C.: 1995).
24. IMF, op. cit. note 23.
25. U.S. Bureau of the Census, as published in Francis Urban and Ray Nightingale, *World Population by Country and Region, 1950-1990, with Projections to 2050* (Washington, D.C.: USDA, Economic Research Service (ERS), 1993); PRB, op. cit. note 23; IMF, op. cit. note 23.
26. USDA, op. cit. note 2.
27. Population from Bureau of the Census, op. cit. note 4; USDA, op. cit. note 2.
28. USDA, ERS, "Food Costs Review 1995," Agricultural Economic Report 729, Washington, D.C., April 1996.
29. U.N. Development Programme, *Human Development Report 1993* (New York: Oxford University Press, 1993).
30. USDA, FAS, *Grain: World Markets and Trade*, Washington, D.C., December 1995.
31. USDA, op. cit. note 2; Egypt population projections from Bureau of the Census, op. cit. note 25.

CHAPTER 2. The Debate

1. Donald O. Mitchell and Merlinda D. Ingco, International Economics Department, *The World Food Outlook* (Washington, D.C.: World Bank, 1993); U.N. Food and Agriculture Organization (FAO), *World Agriculture: Towards 2010* (New York: John Wiley & Sons, 1995).
2. Figure 2-1 from U.S. Department of Agriculture (USDA), "World Grain Database" (unpublished printouts), Washington, D.C., 1991, and from USDA, "Production, Supply, and Distribution" (electronic database), Washington, D.C., February 1996; 1996 estimates from USDA, Foreign Agricultural Service (FAS), *Grain: World Markets and Trade*, Washington, D.C., June 1996.
3. USDA (electronic database), op. cit. note 2; K.F. Isherwood and K.G. Soh, "Short Term Prospects for World Agriculture and Fertilizer Use," presented at 21st Enlarged Council Meeting,

International Fertilizer Industry Association, Paris, November 15-17, 1995.

4. Figure 2-2 from USDA (unpublished printouts), op. cit. note 2, from USDA (electronic database), op. cit. note 2, from USDA, "World Agricultural Supply and Demand Estimates," Washington, D.C, January 1996, and from USDA, FAS, *Grain: World Markets and Trade,* Washington, D.C, January 1996.

5. Figure 2-3 from U.S. Bureau of the Census, Center for International Research, Suitland, Md., private communication, January 23, 1996, and from USDA (electronic database), op. cit. note 2; "Grain Prices Continue to Climb; Official Urges Calmer Trading," *New York Times,* April 26, 1996; "Prices for Wheat and Corn Drop on Hopes for Improved Harvests," *New York Times,* April 30, 1996.

6. Bureau of the Census, op. cit. note 5.

7. FAO, *FAO Production Yearbook 1992* (Rome: 1993); FAO, *FAO Production Yearbook 1991* (Rome: 1992); FAO, *1945-1985 World Crop and Livestock Statistics* (Rome: 1987).

8. Figure 2-4 from FAO, *1945-1985 Crop and Livestock,* op. cit. note 7, from FAO, *FAO Production Yearbooks 1988-1991* (Rome: 1989-92), from USDA, FAS, *Livestock and Poultry: World Markets and Trade,* Washington, D.C., October 1995, and from FAO, *Yearbook of Fishery Statistics: Catches and Landings* (Rome: various years).

9. Figure 2-5 from FAO, *1945-1985 Crop and Livestock,* op. cit. note 7, from FAO, *FAO Production Yearbooks,* op. cit. note 8, from USDA, op. cit. note 8, from FAO, *Yearbook of Fishery Statistics,* op. cit. note 8, and from Bureau of the Census, op. cit. note 5; information on fisheries decline based on statistical data in the FAO fisheries database, FISHSTAT-PC, FAO Fisheries Statistics Division, Rome, 1994.

10. Bureau of the Census, op. cit. note 5.

11. FAO, *Yearbook of Fishery Statistics,* op. cit. note 8; USDA (electronic database), op. cit. note 2.

12. Mitchell and Ingco, op. cit. note 1; FAO, op. cit. note 1.

13. Figure 2-6 from Mitchell and Ingco, op. cit. note 1, from FAO, op. cit. note 1, from USDA (unpublished printouts), op. cit. note 2, and from USDA (electronic database), op. cit. note 2.

14. Mitchell and Ingco, op. cit. note 1.

15. Ibid.

16. FAO, *Yearbook of Fishery Statistics,* op. cit. note 8.

17. "Chinese Roads Paved with Gold," *Financial Times,* November 23, 1994.

18. Christopher Flavin, "Facing Up to the Risks of Climate Change," in Lester R. Brown et al., *State of the World 1996* (New

York: W.W. Norton & Company, 1996).
19. Grain prices from Mitchell and Ingco, op. cit. note 1.
20. "Grain Prices Could Double by 2010," *Kyodo News*, December 25, 1995; "Big Rise in Grain Price Predicted," *China Daily*, December 26, 1995.

CHAPTER 3. Demand for Grain Soaring

1. Joel E. Cohen, "How Many People Can Earth Hold?" *Discover*, November 1992; Figure 3-1 from U.S. Bureau of the Census, Center for International Research, Suitland, Md., private communication, January 23, 1996.
2. Population Reference Bureau (PRB), *1995 World Population Data Sheet* (Washington, D.C.: 1995).
3. International Monetary Fund (IMF), *World Economic Outlook, October 1995* (Washington, D.C.: 1995); population information from PRB, op. cit. note 2; population projection from U.S. Bureau of the Census, as published in Francis Urban and Ray Nightingale, *World Population by Country and Region, 1950-1990, with Projections to 2050* (Washington, D.C.: U.S. Department of Agriculture (USDA), Economic Research Service (ERS), 1993).
4. IMF, op. cit. note 3; population from PRB, op. cit. note 2.
5. U.N. Food and Agriculture Organization (FAO), *FAO Production Yearbook 1994* (Rome: 1995); USDA, "Production, Supply, and Distribution" (electronic database), Washington, D.C., February 1996; PRB, op. cit. note 2; Table 3-1 from USDA, Foreign Agricultural Service (FAS), *Livestock and Poultry: World Markets and Trade*, Washington, D.C., October 1995.
6. FAO, op. cit. note 5; USDA, *Livestock and Poultry*, op. cit. note 5.
7. FAO, op. cit. note 5; poultry information for India from USDA, *Livestock and Poultry*, op. cit. note 5.
8. FAO, *1945-1985 World Crop and Livestock Statistics* (Rome: 1987).
9. Figure 3-2 from ibid., from FAO, *FAO Production Yearbooks 1988-1991* (Rome: 1989-92), and from USDA, *Livestock and Poultry*, op. cit. note 5; population information from Bureau of the Census, op. cit. note 1.
10. Alan Durning, *How Much is Enough? The Consumer Society and the Future of the Earth* (New York: W.W. Norton & Company, 1992); USDA (electronic database), op. cit. note 5; USDA, *Livestock and Poultry*, op. cit. note 5; USDA, FAS, *Livestock and Poultry: World Markets and Trade*, Washington, D.C., 1991.
11. Durning, op. cit. note 10.
12. Ibid.; USDA, FAS, "World Agricultural Production," Washington, D.C., September 1991; USDA, FAS, "World

Agricultural Production," Washington, D.C., October 1995; USDA, *Livestock and Poultry*, op. cit. note 5.

13. USDA, *Livestock and Poultry*, op. cit. note 5; USDA, FAS, "World Agricultural Production," Washington, D.C., August 1995.

14. IMF, op cit. note 3.

15. USDA (electronic database), op. cit. note 5.

16. Corn price from USDA, FAS, *Grain: World Markets and Trade*, Washington, D.C., December 1995; Figure 3-3 from USDA (electronic database), op. cit. note 5.

17. USDA (electronic database), op. cit. note 5.

18. USDA, op. cit. note 16; Ray Goldberg, Moffett Professor of Agriculture and Business, Harvard Graduate School of Business Administration, presentation at "Feeding China: Today and Into the 21st Century," Harvard University, Cambridge, Mass., March 1, 1996.

19. Grain-to-beef conversion ratio based on Allen Baker, Feed Situation and Outlook Staff, ERS, USDA, Washington, D.C., private communication, April 27, 1992; grain-to-pork conversion from Leland Southard, Livestock and Poultry Situation and Outlook Staff, ERS, USDA, Washington, D.C., private communication, April 27, 1992; cheese and egg conversion ratios from Alan B. Durning and Holly B. Brough, *Taking Stock: Animal Farming and the Environment*, Worldwatch Paper 103 (Washington, D.C.: Worldwatch Institute, July 1991); grain-to-poultry ratio derived from Robert V. Bishop et al., *The World Poultry Market—Government Intervention and Multilateral Policy Reform* (Washington, D.C.: USDA, 1990); grain-to-fish ratio from Ross Garnaut and Guonan Ma, East Asian Analytical Unit, Department of Foreign Affairs and Trade, *Grain in China* (Canberra: Australian Government Publishing Service, 1992).

20. Figure 3-4 from FAO, *FAO Production Yearbooks* (Rome: various years).

21. Seth Faison, "A Surprisingly Bitter Brew," *New York Times*, December 27, 1995; *World Drink Trends 1995* (United Kingdom: Produktschap Voor Gedistilleerde Dranken, 1995).

22. USDA (electronic database), op. cit. note 5.

CHAPTER 4. Land Hunger Intensifying

1. This chapter relies heavily on the grain harvested area as an indicator of land availability for two reasons. First, two thirds of the world's cropland is used to produce grain. And second, the area of grain harvested is a much more precise indicator than overall

cropland area. The latter is defined differently by national governments, making it difficult to aggregate data at the global level.

2. U.N. Food and Agriculture Organization (FAO), *The State of Food and Agriculture 1995* (Rome: 1995); Figure 4-1 from U.S. Department of Agriculture (USDA), "Production, Supply, and Distribution" (electronic database), Washington, D.C., February 1996, and from USDA, "World Grain Database" (unpublished printout), Washington, D.C., 1991.

3. USDA (electronic database), op. cit. note 2.

4. Ibid.; Kazak studies from FAO, op. cit. note 2.

5. USDA, Economic Research Service (ERS), *Agricultural Resources: Cropland, Water and Conservation Situation and Outlook Report*, Washington, D.C., September 1991.

6. 1950-59 data from USDA (unpublished printout), op. cit. note 2; other data from USDA (electronic database), op. cit. note 2.

7. Wang Rong, "Food Before Golf on Southern Land," *China Daily*, January 25, 1995.

8. USDA (electronic database), op. cit. note 2.

9. Scott Thompson, "The Evolving Grain Markets in Southeast Asia," in USDA, Foreign Agricultural Service (FAS), *Grain:World Markets and Trade*, Washington, D.C., June 1995; "Ford Avoids Hanoi Farmland Ban," *The Japan Times*, July 25, 1995.

10. "South Korean Golf Course Exempted from Decree on Rice Fields," *New Frontiers*, June 1995.

11. Farmland Mapping and Monitoring Program, California Department of Conservation, *Farmland Conversion Report 1990 to 1992* (Sacramento, Calif.: 1994); Carey Goldberg, "Alarm Bells Sounding as Suburbs Gobble Up California's Richest Farmland," *New York Times*, June 20, 1996.

12. USDA (electronic database), op. cit. note 2.

13. Sandra Postel, *Last Oasis: Facing Water Scarcity* (New York: W.W. Norton & Company, 1992).

14. Figure 4-2 based on "Chinese Roads Paved with Gold," *Financial Times*, November 23, 1994.

15. N. Vasuki Rao, "World's Top Automakers On the Road to India," *Journal of Commerce*, February 26, 1996; U.S. Bureau of the Census, as published in Francis Urban and Ray Nightingale, *World Population by Country and Region, 1950-1990, with Projections to 2050* (Washington, D.C.: USDA, ERS, 1993).

16. Patrick E. Tyler, "China's Transport Gridlock: Cars vs. Mass Transit," *New York Times*, May 4, 1996.

17. Figure 4-3 from FAO, *FAO Production Yearbook* (Rome: various years), and from USDA (electronic database), op. cit. note 2.

18. Table 4-1 from USDA, ERS, *China: Situation and Outlook Report*,

Washington, D.C., August 1994.

19. USDA (electronic database), op. cit. note 2; Bureau of the Census, op. cit. note 15.

20. K.F. Isherwood and K.G. Soh, "Short Term Prospects for World Agriculture and Fertilizer Use," presented at 21st Enlarged Council Meeting, International Fertilizer Industry Association, Paris, November 15-17, 1995; USDA, FAS, "World Agricultural Production," Washington, D.C., October 1995; USDA, FAS, *Grain: World Markets and Trade*, Washington, D.C., December 1995.

21. USDA, ERS, *Agricultural Resources Inputs: Situation and Outlook Report*, Washington, D.C., October 1993.

22. Figure 4-4 from USDA (electronic database), op. cit. note 2, from USDA (unpublished printout), op. cit. note 2, and from USDA, FAS, *Grain: World Markets and Trade*, January 1996, with population from U.S. Bureau of the Census, Center for International Research, Suitland, Md., private communication, January 23, 1996, and from Bureau of the Census, op. cit. note 15; projections calculated using 1990 grain harvested area.

CHAPTER 5. Water Scarcity Spreading

1. U.N. Food and Agriculture Organization (FAO), *FAO Production Yearbook 1993* (Rome: 1994); Bill Quinby, Economic Research Service (ERS), U.S. Department of Agriculture (USDA), Washington, D.C., private communication, January 24, 1996.

2. Sandra Postel, *Last Oasis: Facing Water Scarcity* (New York: W.W. Norton & Company, 1992).

3. Ibid.; Figure 5-1 from FAO, *FAO Production Yearbook* (Rome: various years), and from Quinby, op. cit. note 1.

4. Irrigated area in Figure 5-2 from FAO, op. cit. note 3, with per capita figures derived from U.S. Bureau of the Census, Center for International Research, Suitland, Md., private communication, January 23, 1996.

5. FAO, op. cit. note 1; Quinby, op. cit. note 1.

6. FAO, *FAO Production Yearbook 1994* (Rome: 1995).

7. W. Hunter Colby et al., *Agricultural Statistics of the People's Republic of China, 1949-90* (Washington, D.C.: USDA, ERS, 1992).

8. Postel, op. cit. note 2, citing M.G. Chandrakanth and Jeff Romm, "Groundwater Depletion in India—Institutional Management Regimes," *Natural Resources Journal*, Summer 1990, and U.S. Geological Survey, *Estimated Water Use of the United States in 1990*

(Washington, D.C.: U.S. Government Printing Office, 1992).

9. Gordon Sloggett and Clifford Dickason, *Ground-Water Mining in the United States* (Washington, D.C.: USDA, ERS, 1986); information on Iran from Gary Gardner, "Preserving Agricultural Resources," in Lester R. Brown et al., *State of the World 1996* (New York: W.W. Norton & Company, 1996).

10. USDA, Natural Resources Conservation Service, *Summary Report: 1992 National Resources Inventory*, Washington, D.C., July 1994, rev. January 1995; Sandra Postel, "Forging a Sustainable Water Strategy," in Brown et al., op. cit. note 9.

11. Gardner, op. cit. note 9; R.P.S. Malik and Paul Faeth, "Rice-Wheat Production in Northwest India," in Paul Faeth, ed., *Agricultural Policy and Sustainability: Case Studies from India, Chile, the Philippines, and the United States* (Washington, D.C.: World Resources Institute, 1993).

12. Professor Chen Yiyu, Chinese Academy of Sciences, Beijing, China, private communication, March 12, 1996.

13. Vaclav Smil, "Environmental Problems in China," East-West Center Special Reports No. 5, East-West Center, Honolulu, Hawaii, April 1996; information on Handan from "Severe Water Shortages Hit Northeastern China," Knight-Ridder News Service, January 5, 1996.

14. Vaclav Smil, *China's Environmental Crisis: An Inquiry Into the Limits of National Development* (Armonk, N.Y.: M.E. Sharpe, 1993).

15. Current and projected population from U.S. Bureau of the Census, as published in Francis Urban and Ray Nightingale, *World Population by Country and Region, 1950-1990, with Projections to 2050* (Washington, D.C.: USDA, ERS, 1993).

16. "CHINA: Commentator on Urgency of Water Conservation," *Environment and World Health*, October 31, 1995; Postel, op. cit. note 2.

17. Gerald Barney and Weishuang Qu, Millennium Institute, Arlington, Va., private communication, May 7, 1996.

18. McVean Trading and Investments, Memphis, Tenn., private communication, May 29, 1996.

19. Postel, op. cit. note 10.

20. Ibid.

21. "Western States Water Conservation: What Are the States Doing? What is the Federal Role?" Environmental and Energy Study Institute Report, Washington, D.C., March 8, 1996.

22. Leonard and Shirley Ann Wiggin, ranchers in Grover, Colo., private communication, December 24, 1994; local officials, Fukuoka, Japan, private communication, July 20, 1995.

23. Patrick E. Tyler, "Huge Water Project Would Supply Beijing By 860-Mile Aqueduct," *New York Times*, July 19, 1994.

24. Patrick E. Tyler, "China's Fickle Rivers: Dry Farms, Needy Industry Bring a Water Crisis," *New York Times*, May 23, 1996.

25. Ibid.

26. "Water in the Middle East," *The Economist*, December 23rd, 1995-January 5th, 1996; population growth rate from Bureau of the Census, op. cit. note 15.

27. Sandra L. Postel, Gretchen C. Daily, and Paul R. Ehrlich, "Human Appropriation of Renewable Fresh Water," *Science*, February 9, 1996.

28. Postel, op. cit. note 10.

29. Postel, op. cit. note 2.

30. John Barham, "Euphrates Power Plant Generates New Tension," *Financial Times*, February 15, 1996; Yuan Shu, "Nations Find Unity in Taming the Mekong," *The WorldPaper*, November 1994.

31. David Seckler, "The New Era of Water Resources Management: From 'Dry' to 'Wet' Water Savings," Consultative Group on International Agricultural Research, Washington, D.C., April 1996.

CHAPTER 6. Rise in Land Productivity Slowing

1. U.S. Department of Agriculture (USDA), "World Grain Database" (unpublished printout), Washington, D.C., 1991; USDA, "Production, Supply, and Distribution" (electronic database), Washington, D.C., February 1996.

2. Figure 6-1 from USDA (electronic database), op. cit. note 1, with production figures for 1950-59 from USDA (unpublished printout), op. cit. note 1.

3. USDA (unpublished printout), op. cit. note 1; USDA (electronic database), op. cit. note 1.

4. USDA (unpublished printout), op. cit. note 1; USDA (electronic database), op. cit. note 1.

5. Table 6-1 based on USDA (unpublished printout), op. cit. note 1; USDA (electronic database), op. cit. note 1.

6. Table 6-2 from USDA (electronic database), op. cit. note 1, with data for 1950-59 from USDA (unpublished printout), op. cit. note 1.

7. U.N. Food and Agriculture Organization (FAO), *Fertilizer Yearbook* (Rome: various years); K.F. Isherwood and K.G. Soh, "Short Term Prospects for World Agriculture and Fertilizer Use," presented at 21st Enlarged Council Meeting, International

Fertilizer Industry Association, Paris, November 15-17, 1995.

8. Figure 6-2 from FAO, op. cit. note 7, and from Isherwood and Soh, op. cit. note 7.

9. FAO, op. cit. note 7; Isherwood and Soh, op. cit. note 7.

10. Figure 6-3 from FAO, *FAO Production Yearbooks* (Rome: various years), from FAO, op. cit. note 7, and from Isherwood and Soh, op. cit. note 7.

11. FAO, *Fertilizer Yearbooks*, op. cit. note 7.

12. Leon Lyles, "Possible Effects of Wind Erosion on Soil Productivity," *Journal of Soil and Water Conservation*, November/December 1975.

13. James J. MacKenzie and Mohamed T. El-Ashry, *Ill Winds: Airborne Pollution's Toll on Trees and Crops* (Washington, D.C.: World Resources Institute, 1988); National Acid Precipitation Assessment Program, *Interim Assessment: The Causes and Effects of Acid Deposition*, Vol. IV (Washington, D.C.: U.S. Government Printing Office, 1987).

14. "Forests, Crops Suffering Ozone Damage," *Dagens Nyheter*, July 5, 1990, as reprinted in *JPRS Report: Environmental Issues*, October 12, 1990; Dr. Jan Cerovsky, "Environmental Status Report 1988/89: Czechoslovakia," in World Conservation Union—IUCN, *Environmental Status Reports: 1988/89, Vol. 1: Czechoslovakia, Hungary, Poland* (Thatcham, U.K.: Thatcham Printers, 1990).

15. USDA (electronic database), op. cit. note 1.

16. Figure 6-4 from Odil Tunali, "Global Temperature Sets New Record," in Lester R. Brown, Christopher Flavin, and Hal Kane, *Vital Signs 1996* (New York: W.W. Norton & Company, 1996).

17. Donald O. Mitchell and Merlinda D. Ingco, International Economics Department, *The World Food Outlook* (Washington, D.C.: World Bank, 1993).

18. USDA (electronic database), op. cit. note 1.

19. Ibid.

20. Lester R. Brown, *Seeds of Change* (New York: Praeger Publishers, 1970); USDA, National Agricultural Statistics Service, *Agricultural Statistics 1994* (Washington, D.C.: U.S. Government Printing Office, 1994).

21. Paul F. Conley, "Precision Farming," *Journal of Commerce*, April 19, 1996.

22. Ibid.

23. "Justus von Liebig" and "Gregor Mendel," *Encyclopaedia Britannica* (Cambridge, Mass.: Encyclopaedia Britannica, Inc., 1976); "History of Agriculture," ibid.; Joseph A. Tainter, *The Collapse of Complex Societies* (New York: Cambridge University

Press, 1988).

24. Donald N. Duvick, "Intensification of Known Technology and Prospects of Breakthroughs in Technology and Future Food Supply," Iowa State University, Johnston, Iowa, February 1994.

25. John Madeley, "Rice—The Next Generation," *Financial Times*, February 18, 1994; USDA (electronic database), op. cit. note 1; population growth from Population Reference Bureau, *1995 World Population Data Sheet* (Washington, D.C.: 1995).

26. Figure 6-5 from FAO, *Fertilizer Yearbooks*, op. cit. note 10, from International Fertilizer Industry Association, *Fertilizer Consumption Report* (Paris: 1992), from USDA (unpublished printout), op. cit. note 1, from USDA (electronic database), op. cit. note 1, and from U.S. Bureau of the Census, Center for International Research, Suitland, Md., private communication, January 23, 1996.

CHAPTER 7. The New Politics of Scarcity

1. Table 7-1 based on references cited in earlier chapters, which contain details of each indicator in the table.

2. Temperature rise since 1979 from James Hansen et al., Goddard Institute for Space Studies Surface Air Temperature Analyses, "Table of Global-Mean Monthly, Annual and Seasonal Land-Ocean Temperature Index, 1950-Present," as posted at http://www.giss.nasa.gov/Data/GISTEMP, January 19, 1996.

3. U.S. Bureau of the Census, Center for International Research, Suitland, Md., private communication, January 23, 1996; Population Reference Bureau (PRB), *1995 World Population Data Sheet* (Washington, D.C.: 1995); U.S. Department of Agriculture (USDA), "Production, Supply, and Distribution" (electronic database), Washington, D.C., February 1996.

4. U.N. Food and Agriculture Organization (FAO), *Yearbook of Fishery Statistics: Catches and Landings* (Rome: various years); USDA, op. cit. note 3.

5. USDA, op. cit. note 3, with population from Bureau of the Census, op. cit. note 3; "Grain Prices Continue to Climb; Official Urges Calmer Trading," *New York Times*, April 26, 1996; "Prices for Wheat and Corn Drop on Hopes for Improved Harvests," *New York Times*, April 30, 1996.

6. Table 7-2 from Lester R. Brown and Hal Kane, *Full House: Reassessing the Earth's Population Carrying Capacity* (New York: W.W. Norton & Company, 1994).

7. Worldwatch estimates based on USDA, "World Grain Database"

(unpublished printout), Washington, D.C., 1991, on USDA, op. cit. note 3, and on FAO, *Trade Yearbook* (Rome: various years); U.S. Bureau of the Census, as published in Francis Urban and Ray Nightingale, *World Population by Country and Region, 1950-1990, with Projections to 2050* (Washington, D.C.: USDA, Economic Research Service, 1993).

8. Bureau of the Census, op. cit. note 7.
9. USDA, op. cit. note 3; Bureau of the Census, op. cit. note 8; PRB, op. cit. note 3.
10. USDA, op. cit. note 3; historical data from USDA, op. cit. note 7, and from FAO, op. cit. note 7; PRB, op. cit. note 3; Bureau of the Census, op. cit. note 7.
11. USDA, op. cit. note 3; historical data from USDA, op. cit. note 7, from FAO, op. cit. note 7; PRB, op. cit. note 3; Bureau of the Census, op. cit. note 7.
12. Figure 7-1 from USDA, op. cit. note 3, and from USDA, op. cit. note 7.
13. "Grain Prices Continue to Climb," op. cit. note 5.
14. USDA, Foreign Agricultural Service (FAS), *Grain: World Markets and Trade*, Washington, D.C., January 1996; population from PRB, op. cit. note 3.
15. "Vietnam to Limit Exports of Rice for Four Months," *Journal of Commerce*, May 19, 1995; information on China from Christopher Goldthwaite, FAS, USDA, Washington, D.C., private communication, April 25, 1995; USDA, FAS, *Grain: World Markets and Trade*, Washington, D.C., December 1995.
16. "Wheat Soars to 15-Year High As Europe Puts Tax on Exports," *New York Times*, December 8, 1995; "EU to Conserve Barley by Curbing Exports," *Journal of Commerce*, January 12, 1996.
17. USDA, op. cit. note 14.
18. Greenpeace quote from William Branigin, "Global Accord Puts Curbs on Fishing," *Washington Post*, August 4, 1995.
19. Sandra Postel, "Forging a Sustainable Water Strategy," in Lester R. Brown et al., *State of the World 1996* (New York: W.W. Norton & Company, 1996).
20. FAO, *Food Outlook*, August/September 1995.
21. Soybean export ban from Lester R. Brown, *By Bread Alone* (New York: Praeger Publishers, 1974).
22. USDA, op. cit. note 3.
23. Ibid.
24. "Vietnam to Limit Exports," op. cit. note 15; "Wheat Soars to 15-Year High," op. cit. note 16; "EU to Conserve Barley," op. cit. note 16.

CHAPTER 8. Responding to the Challenge

1. Population Reference Bureau (PRB), *1995 World Population Data Sheet* (Washington, D.C.: 1995).
2. Irene V. Taeuber, *The Population of Japan* (Princeton, N.J.: Princeton University Press, 1958).
3. U.S. Bureau of the Census, as published in Francis Urban and Ray Nightingale, *World Population by Country and Region, 1950-1990, with Projections to 2050* (Washington, D.C.: U.S. Department of Agriculture (USDA), Economic Research Service, 1993).
4. Ayatollah quote from "Fewer Means Better," *The Economist*, August 8, 1995; Anthony Shadid, "Population—Iran," Associated Press—US & World Wire Service, February 6, 1995; Bureau of the Census, op. cit. note 3; Arjun Adlakha, U.S. Bureau of the Census, International Division, Suitland, Md., private communication, May 31, 1996.
5. Akbar Aghajanian, "A New Direction in Population Policy and Family Planning in the Islamic Republic of Iran," *Asia-Pacific Population Journal*, March 1995; Carl Haub, PRB, Washington, D.C., private communication, May 31, 1996.
6. "Fewer Means Better," op. cit. note 4.
7. Ibid.; Aghajanian, op. cit. note 5.
8. Unmet need from U.N. General Assembly, "Draft Programme of Action of the International Conference on Population and Development" (draft), New York, April 1994.
9. See the section on "Energy Trends" in Lester R. Brown, Christopher Flavin, and Hal Kane, *Vital Signs 1996* (New York: W.W. Norton & Company, 1996).
10. California wind farm potential from Paul Gipe, Gipe and Associates, Tehachapi, Calif., private communication and printout, April 7, 1994; potential in the three U.S. states from D.L. Elliott, L.L. Windell, and G.L. Gower, *An Assessment of the Available Windy Land Area and Wind Energy Potential in the Contiguous United States* (Richland, Wash.: Pacific Northwest Laboratory, 1991); Andrew Garrad, *Wind Energy in Europe: Time for Action* (Rome: European Wind Energy Association, 1991); information on China from Christopher Flavin and Nicholas Lenssen, *Power Surge: Guide to the Coming Energy Revolution* (New York: W.W. Norton & Company, 1994); hydropower estimate from United Nations, *1990 Energy Statistics Yearbook* (New York: 1992).
11. Odil Tunali, "Solar Cell Shipments Jump," in Brown, Flavin, and Kane, op. cit. note 9.

12. Flavin and Lenssen, op. cit. note 10.
13. Gerhard A. Berz, Munich Reinsurance Company, Munich, Germany, private communication, September 1, 1995.
14. "Many of Last Year's Heat Deaths Were Preventable, Study Says," *Washington Post*, April 12, 1996; USDA, "Production, Supply and Distribution" (electronic database), Washington, D.C., February 1996.
15. Patrick E. Tyler, "China's Transport Gridlock: Cars vs. Mass Transit," *New York Times*, May 4, 1996.
16. U.N. Food and Agriculture Organization (FAO), *Yield Response to Water* (Rome: 1979).
17. Gary Gardner, "Preserving Agricultural Resources," in Lester R. Brown et al., *State of the World 1996* (New York: W.W. Norton & Company, 1996).
18. K.F. Isherwood and K.G. Soh, "Short Term Prospects for World Agriculture and Fertilizer Use," presented at 21st Enlarged Council Meeting, International Fertilizer Industry Association, Paris, November 15-17, 1995; USDA, Foreign Agricultural Service (FAS), "World Agricultual Production," Washington, D.C., October 1995; USDA, FAS, *Grain: World Markets and Trade*, Washington, D.C., December 1995; PRB, op. cit. note 1.
19. USDA, FAS, *Grain: World Markets and Trade*, Washington, D.C., March 1995; PRB, op. cit note 1.
20. USDA, op. cit. note 14; PRB, op. cit. note 1; Robert Greene, "Low Grain Supplies Put Pressure on Livestock," Associated Press, April 12, 1996.
21. FAO, *FAO Production Yearbook 1994* (Rome: 1995); USDA, op. cit. note 14; PRB, op. cit. note 1.
22. FAO, op. cit. note 21; USDA, op. cit. note 14; PRB, op. cit. note 1.
23. USDA, op. cit. note 14.
24. Ibid.
25. FAO, *FAO Production Yearbook 1990* (Rome: 1991); life expectancy from PRB, op. cit. note 1.
26. USDA, op. cit. note 14; PRB, op. cit. note 1.
27. Peter Kostmayer, Executive Director, Zero Population Growth, Washington, D.C., private communication, April 24, 1996.

Index

ABOUT THE AUTHOR

LESTER R. BROWN is President of Worldwatch Institute, a private, nonprofit environmental research organization in Washington, D.C. He is the recipient of a MacArthur Foundation "genius award," the United Nations' 1989 environmental prize, and the Asahi Glass Foundation's Blue Planet Prize, and he holds a string of honorary degrees from universities around the world. The Library of Congress has requested Mr. Brown's personal papers and manuscripts, recognizing his role and that of the Institute that he heads in shaping the global environmental movement. Before founding Worldwatch, he was Administrator of the U.S Department of Agriculture's International Agricultural Development Service and Advisor to the Secretary. He holds degrees from Rutgers University, the University of Maryland, and Harvard University. Mr. Brown started his career as a farmer, growing tomatoes in southern New Jersey.